NOAA'S EDUCATION PROGRAM

Review and Critique

Committee for the Review of the NOAA Education Program

John W. Farrington and Michael A. Feder, *Editors*

Board on Science Education

Division of Behavioral and Social Sciences and Education

NATIONAL RESEARCH COUNCIL
OF THE NATIONAL ACADEMIES

THE NATIONAL ACADEMIES PRESS
Washington, D.C.
www.nap.edu

THE NATIONAL ACADEMIES PRESS 500 Fifth Street, N.W. Washington, DC 20001

NOTICE: The project that is the subject of this report was approved by the Governing Board of the National Research Council, whose members are drawn from the councils of the National Academy of Sciences, the National Academy of Engineering, and the Institute of Medicine. The members of the committee responsible for the report were chosen for their special competences and with regard for appropriate balance.

This study was supported by Contract No. DG133R07CN0261 between the National Academy of Sciences and the National Oceanic and Atmospheric Administration. Any opinions, findings, conclusions, or recommendations expressed in this publication are those of the authors and do not necessarily reflect the views of the National Oceanic and Atmospheric Administration.

International Standard Book Number-13: 978-0-309-15123-8
International Standard Book Number-10: 0-309-15123-6

Additional copies of this report are available from National Academies Press, 500 Fifth Street, N.W., Lockbox 285, Washington, DC 20055; (800) 624-6242 or (202) 334-3313 (in the Washington metropolitan area); Internet, http://www.nap.edu.

Printed in the United States of America

Suggested citation: National Research Council. (2010). *NOAA's Education Program: Review and Critique*. Committee for the Review of the NOAA Education Program. J.W. Farrington and M.A. Feder, Editors. Board on Science Education, Division of Behavioral and Social Sciences and Education. Washington, DC: The National Academies Press.

THE NATIONAL ACADEMIES

Advisers to the Nation on Science, Engineering, and Medicine

The **National Academy of Sciences** is a private, nonprofit, self-perpetuating society of distinguished scholars engaged in scientific and engineering research, dedicated to the furtherance of science and technology and to their use for the general welfare. Upon the authority of the charter granted to it by the Congress in 1863, the Academy has a mandate that requires it to advise the federal government on scientific and technical matters. Dr. Ralph J. Cicerone is president of the National Academy of Sciences.

The **National Academy of Engineering** was established in 1964, under the charter of the National Academy of Sciences, as a parallel organization of outstanding engineers. It is autonomous in its administration and in the selection of its members, sharing with the National Academy of Sciences the responsibility for advising the federal government. The National Academy of Engineering also sponsors engineering programs aimed at meeting national needs, encourages education and research, and recognizes the superior achievements of engineers. Dr. Charles M. Vest is president of the National Academy of Engineering.

The **Institute of Medicine** was established in 1970 by the National Academy of Sciences to secure the services of eminent members of appropriate professions in the examination of policy matters pertaining to the health of the public. The Institute acts under the responsibility given to the National Academy of Sciences by its congressional charter to be an adviser to the federal government and, upon its own initiative, to identify issues of medical care, research, and education. Dr. Harvey V. Fineberg is president of the Institute of Medicine.

The **National Research Council** was organized by the National Academy of Sciences in 1916 to associate the broad community of science and technology with the Academy's purposes of furthering knowledge and advising the federal government. Functioning in accordance with general policies determined by the Academy, the Council has become the principal operating agency of both the National Academy of Sciences and the National Academy of Engineering in providing services to the government, the public, and the scientific and engineering communities. The Council is administered jointly by both Academies and the Institute of Medicine. Dr. Ralph J. Cicerone and Dr. Charles M. Vest are chair and vice chair, respectively, of the National Research Council.

www.national-academies.org

COMMITTEE FOR THE REVIEW OF THE NOAA EDUCATION PROGRAM

JOHN W. FARRINGTON (*Chair*), Woods Hole Oceanographic Institution, and University of Massachusetts Dartmouth
JAMES M. COLEMAN, Coastal Studies Institute, Louisiana State University and Agricultural and Mechanical College
JANET HUSTLER, Partnership for Student Success in Science (PS^3), Synopsys, Inc.
KIM A. KASTENS, Lamont-Doherty Earth Observatory, Columbia University
GORDON KINGSLEY, School of Public Policy, Georgia Institute of Technology
KEVIN KLOESEL, College of Atmospheric and Geographic Sciences, University of Oklahoma
FRANCES LAWRENZ, Department of Educational Psychology, University of Minnesota
GEORGE L. MATSUMOTO, Monterey Bay Aquarium Research Institute
BRETT D. MOULDING, Utah Partnership for Effective Science Teaching and Learning
FRANK E. MULLER-KARGER, College of Marine Science, University of South Florida
LAURA MURRAY, Center for Environmental Science, University of Maryland
RAJUL PANDYA, University Cooperation for Atmospheric Research, Boulder, CO
CRAIG STRANG, Lawrence Hall of Science, University of California, Berkeley
CLARICE YENTSCH, Oceanographic Center, Nova Southeastern University

MICHAEL A. FEDER, *Study Director*
CLAUDIA MENGELT, *Senior Program Officer*
REID SCHWEBACH, *Program Officer*
HEIDI A. SCHWEINGRUBER, *Deputy Director,* Board on Science Education
REBECCA KRONE, *Program Associate*
PATRICIA HARVEY, *Senior Program Assistant* (until July 2009)
WUNIKA MUKAN, *Program Assistant* (until December 2009)
MIRIAM QUINTAL, *Christine Mirzayan Science and Technology Policy Graduate Fellow* (until June 2009)

*Until May 2009.

Acknowledgments

The committee and staff thank the many individuals and organizations who assisted us in our work, without whom this study could not have been completed. First, we acknowledge the support of staff in the Office of Education of the National Oceanic and Atmospheric Administration (NOAA), who made themselves readily available to us and provided detailed information about the agency's education activities and its overarching priorities. They were quick to respond to requests for information, contacted other NOAA staff to help field requests, and were persistent in obtaining the information requested by the committee. Louisa Koch, director, and Christos Michalopoulos, assistant director, were especially helpful.

Individually and collectively, committee members benefited from discussions and presentations by the many individuals who participated in our four fact-finding meetings. We are grateful to each of the presenters: Glen Alexander, education coordinator, Padilla Bay National Estuarine Research Reserve; Eric Bolt, warning coordination meteorologist, National Weather Service; Leon Cammen, director, National Sea Grant College Program; Clarice Fackler, national education liaison, Office of National Marine Sanctuaries; Ron Gird, outreach program manager, National Weather Service; Jennifer Hammond, director, Teacher at Sea Program; Molly Harrison, national education coordinator, National Marine Fisheries Service; Atziri Ibanez, national education coordinator, National Estuarine Research Reserve System; Paula Keener-Chavis, director of education programs, Ocean Exploration and Research Program; Louisa Koch, Office of Education; Michiko Martin, national education coordinator, Office of National Marine Sanctuaries; Laurie McGilvray, chief, National Estuarine Research Reserves; Christos Michalopoulos, Office of Education; Seaberry Nachbar, program manager, B-WET California; Frank Niepold, climate education coordinator; Jacqueline Rousseau, director, Educational Partnership Program; Sharon Walker, education director, National Sea Grant College Program; and Marci Wulff, Corals Program specialist, Coral Reef Conservation Program.

The committee also benefited from the contributions of NOAA staff and staff from partnering agencies who participated in our two site visits, including Tom Ackerman, director of teaching and training and student leadership, Chesapeake Bay Foundation; Jamie Baxter, program director, Chesapeake Bay Trust; Lisa Emanuelson, volunteer monitoring coordinator, Monterey Bay National Marine Sanctuary; Ellen Fondiler, project manager, Hilton Bialek Biological Science Habitat; Dawn Hayes, education and outreach coordinator, Monterey Bay National Marine Sanctuary; Doug Levin, habitat specialist/education coordinator, Chesapeake Bay Office; Sacha Lozano, MERITO program coordinator, Monterey Bay National Marine Sanctuary; Paul Michel, sanctuary superintendent, Monterey Bay National Marine Sanctuary; Seaberry Nachbar, program manager, California B-WET; Kenton Parker, education coordinator, Elkhorn Slough National Estuarine Research Reserve; Kevin Schabow, education specialist, Chesapeake Bay Office; Shannon Sprague, environmental literacy manager, Chesapeake Bay Office; and Elena Takaki, program manager, Maryland Department of Natural Resources.

The committee also benefited from presentations by other experts knowledgeable about the ocean, atmosphere, climate, and environmental education policy and federal interagency groups. Thanks to Daniel Barstow, president, Challenger Center for Space Science Education; Jim Brey, education program director, American Meteorological Society; Sue Cook, education director, Consortium for Ocean Leadership; Roberta Johnson, director, University Corporation for Atmospheric Research; Jill Karsten, cochair, U.S. Climate Change Science Program Education Interagency Working Group; Gerry Lieberman, president, State Environmental Education Roundtable; Frank Niepold, cochair, U.S. Climate Change Science Program Education Interagency Working Group; Jeffrey Reutter, advisory panel member, Ocean Research Resources Advisory Panel; Lisa Rom, cochair, National Science Technology Council Joint Subcommittee on Ocean Science and Technology; and Jill Sanders, president, National Association of Marine Laboratories.

The committee is also grateful to the panel of experts that made presentations on issues of diversity and broadening participation in fields critical to NOAA's mission. Thanks to Deidra Gibson, assistant professor,

Hampton University; Reza Khanbilvardi, professor, City University of New York; Roger Levin, managing research scientist, American Institutes for Research; Ramon Lopez, professor, University of Texas at Arlington; Eric Riggs, associate professor, Purdue University; and Larry Robinson, professor, Florida A&M University.

The committee benefited from the contributions of the authors of four papers whose work informed this report. Ann Brackett, an independent consultant, reviewed several external evaluations of K-12 education projects in NOAA and wrote a summary and critique. Bill Clune, Voss-Bascom professor at the Wisconsin Center for Education Research, applied a common logic model across NOAA's education programs and discussed inefficiencies, redundancies, and implementation challenges in each component of the general model as applied to NOAA education programs. Roger Levin, managing research scientist at the American Institutes for Research; Raquel Gonzalez, doctoral student at the University of Maryland, College Park; and Carmen Martínez-Sussman, doctoral student at the University of California, Santa Cruz, synthesized the research on issues of diversity and broadening participation in fields critical to NOAA's mission. Lynn Tran, specialist at the Lawrence Hall of Science, synthesized the research on the teaching and learning of ocean, climate, and atmospheric science.

Many individuals at the National Research Council (NRC) assisted the committee. The Ocean Studies Board and Board on Atmospheric Science and Climate of the Division on Earth and Life Studies provided essential guidance regarding the composition of the study committee. From the Division of Behavioral and Social Sciences and Education, Heidi Schweingruber and Martin Storksdieck offered valuable suggestions at our committee meetings, as well as providing helpful comments on drafts of the report. We thank Kirsten Sampson Snyder, who shepherded the report through the NRC review process, Christine McShane, who edited the draft report, and Yvonne Wise for processing the report through final production. We are grateful to Patricia Harvey, who arranged logistics for the first four committee meetings. Finally, we thank Wunika Mukan for her able assistance in arranging the final committee meeting and in preparing numerous drafts and revisions of the report.

This report has been reviewed in draft form by individuals chosen for their diverse perspectives and technical expertise, in accordance with procedures approved by the NRC's Report Review Committee. The purpose of this independent review is to provide candid and critical comments that will assist the institution in making its published report as sound as possible and to ensure that the report meets institutional standards for objectivity, evidence, and responsiveness to the study charge. The review comments and draft manuscript remain confidential to protect the integrity of the deliberative process. We thank the following individuals for their review of this report: Janet Carlson, Executive Director's Office, Biological Sciences Curriculum Study, Colorado Springs, CO; Inés Cifuentes, Education and Career Services, American Geophysical Union; Frank Kudrna, Director's Office, Kudrna and Associates; Cathryn A Manduca, Science Education Resource Center, Carleton College; Richard A. McCray, Department of Astrophysics, University of Colorado, Boulder; Vera Michalchik, Center for Technology in Learning, SRI International; William S. Spitzer, Programs, Exhibits, and Planning, New England Aquarium; Elizabeth K. Stage, Lawrence Hall of Science, University of California, Berkeley; and Quinton L. Williams, Department of Physics, Atmospheric Science and Geoscience, Jackson State University.

Although the reviewers listed above provided many constructive comments and suggestions, they were not asked to endorse the conclusions or recommendations nor did they see the final draft of the report before its release. Michael E. Martinez, Department of Education, University of California, Irvine, and W. Carl Lineberger, Department of Chemistry, University of Colorado, Boulder, oversaw the review of this report. Appointed by the NRC, they were responsible for making certain that an independent examination of this report was carried out in accordance with institutional procedures and that all review comments were carefully considered. Responsibility for the final content of this report rests entirely with the author and the institution.

Contents

Tables, Figures, and Boxes

TABLES

FIGURES

BOXES

Summary

The National Oceanic and Atmospheric Administration (NOAA) is responsible for understanding and predicting changes in the Earth's environment and conserving and managing coastal and marine resources to meet the nation's economic, social, and environmental needs. Since it was created in 1970, the agency has supported education projects that cover a range of topics related to the agency's scientific and stewardship mission, including oceanic, atmospheric, climate, and environmental sciences. Given human dependence on the Earth for health, well-being, and economic growth, the importance of these interconnected fields and environmental stewardship cannot be overstated.

Education efforts at NOAA are distributed across a range of internal offices. Some of them have long had mandates to engage in education activities, but it was not until 2007 that NOAA received an agencywide mandate for education through the America Creating Opportunities to Meaningfully Promote Excellence in Technology, Education, and Science (COMPETES) Act. The act calls for NOAA to support and coordinate formal and informal educational activities to enhance public awareness and understanding of issues related to its mission. The act also requires that NOAA develop a 20-year education plan, to be reevaluated and updated every 5 years.

The Committee for the Review of the NOAA Education Program was established by the National Research Council (NRC) to take stock of the existing education portfolio and review the education strategic plan mandated by the America COMPETES Act. The committee was specifically asked to comment on:

1. NOAA's role in education,
2. its education goals and outcomes,
3. the composition and management of its education portfolio,
4. its education evaluation practice, and
5. the impact of its education efforts.

The committee followed an iterative process of gathering information, analyzing and deliberating on it, identifying gaps and questions, gathering additional information to fill these gaps, and carrying out further analysis. A contract between NOAA and NRC determined time and resources available for the study and constrained the scope of the committee's review to existing documentation, site visits, testimony from NOAA staff, and commissioned papers. The resulting report provides a summary of the national education context for NOAA's role in education (Chapter 2) and of the education strategic plan and its strengths and weaknesses (Chapter 3). It also describes the individual education projects (Chapter 4). In Chapter 5, the education evaluation approach of the agency is described, and suggestions for improving the process are provided. The final chapter presents the committee's conclusions and recommendations.

NOAA'S ROLE IN EDUCATION

The national need to educate the public about the ocean, coastal resources, atmosphere, and climate and to support workforce development in related fields is well established. The federal government role in addressing these needs as part of the national effort is also widely accepted.

NOAA's role in education has been recognized for approximately 30 years, as evidenced by the mandates to engage in education activities given to individual operating branches and programs, and more recently by the America COMPETES Act. The agency has a broad mandate to engage in and coordinate education and stewardship initiatives related to ocean, Great Lakes, climate, and atmospheric science, as well as other fields related to its mission. NOAA must fulfill these responsibilities in the context of a national effort, implemented at state and local levels. The agency must use formal and informal learning environments to improve learning and understanding of science, technology, engineering, and mathematics (STEM) and to advance environmental education.

Although NOAA is unique among federal agencies in its focus on stewardship and on ocean, coastal, Great Lakes, atmospheric, and climate science, its mission overlaps with and complements the missions of other federal agencies. Many federal agencies, institutions of higher education, and private and nonprofit organizations have additional resources that help improve the nation's understanding and interest in the relevant sciences and that help develop strategies to care for the environment. However, coordination of these activities in a cohesive way that leverages the unique assets of each federal agency, as well as the formidable infrastructure and capabilities outside the federal government, has proven to be a challenge.

NOAA can contribute to national education efforts through a variety of programs and assets, including modern and groundbreaking technologies and discoveries; research equipment; data sets; technical staff, including scientists, engineers, and researchers; stewardship and management of natural resources; specialized education expertise; partnerships; and connections to local, regional, national, and international stakeholders and natural resource managers. In addition, NOAA is one of the key federal agencies engaged in management and stewardship of the coasts and oceans. These natural environments can support important educational opportunities and provide the agency with connections to the surrounding communities and organizations concerned with environmental issues.

NOAA's role in education is shaped by the distributed nature of its education efforts across the five line offices and the Office of Education, the small number of agency staff involved in education, and its small education budget. Because of their diverse missions, the line offices (some of which have individual education mandates) and the Office of Education can act independently and sometimes even in competition with each other. The majority of education programs are usually implemented by an individual or a small team at a particular location. And NOAA's education budget is relatively small in comparison to that of other federal agencies engaged in STEM education, such as the U.S. Department of Education, the National Science Foundation, the National Aeronautics and Space Administration, and the U.S. Department of Energy.

Limited education resources and the inherently global nature of NOAA's mission make strategic partnerships necessary in order for the agency to accomplish its ambitious goals. Clear education goals, planning, and strategic use of resources are critical aspects for effective partnerships.

NOAA can play a supporting role in state and local education systems and a leadership role in federal STEM education endeavors specific to oceanic, coastal, Great Lakes, atmospheric, and climate sciences. Such efforts will be most productive if they align with local education needs and national education standards, because education activities and products that do not consider the needs of the potential audiences are less likely to be successful.

Recommendations Regarding NOAA's Role in Education

Recommendation I.1: NOAA should fulfill its role in education through the use of:

- agency and external expertise in science, engineering, technology, and education; cutting-edge scientific research and exploration activities; internationally collected datasets; and advances in technology and engineering;
- place-based assets that directly connect local issues to national and global science and stewardship issues: marine sanctuaries, estuarine research reserves, fisheries activities, and other natural resources protected and managed by federal, state, and local entities;
- partnerships with local and state education infrastructure, academic institutions, government agencies, business and industry, and private-sector and nonprofit organizations; and
- the agency's global science and international partnerships.

Recommendation I.2: In order to adequately address the mismatch between its available resources and its ambitious education agenda, NOAA should better align and deploy its resources. This may require the termination of certain activities and programs that, based on appropriate evaluation, do not directly and effectively contribute to its education and stewardship goals.

Recommendation I.3: Within the constraints of NOAA's mandates in education, the agency should continually evaluate where it leads, collaborates, follows, or declines to participate in partnerships with others. These decisions should be guided by consideration of the agency's role, assets, resources, and priorities in education and the strengths and missions of other agencies, institutions, and organizations engaged in education.

EDUCATION GOALS AND OUTCOMES

The NOAA education strategic plan for 2009-2029, developed by its recently formed Education Council, provides goals, outcomes, and a framework to organize a large set of individual education activities into a coherent portfolio. The plan outlines two goals: (1) to advance the environmental literacy of the nation, and (2) to promote a diverse workforce in oceanic, coastal, Great Lakes, atmospheric, and climate sciences. At this time, NOAA is developing a strategic implementation plan to specify how it will accomplish these goals. The strategic plan lists six outcomes under the goal to improve environmental literacy and three outcomes under the goal to promote a diverse workforce.

The plan has multiple strengths. It includes appropriate goals of supporting environmental literacy and workforce development and stresses the need for partnerships with appropriate agencies, institutions, and organizations. The plan also illustrates a commitment to developing education programs informed by evidence about effective practices and contributing to this body of knowledge. In addition, there is an emphasis on the use of ocean, coastal, and other place-based resources as unique and valuable assets for learning. Overall, the 2009-2029 education strategic plan is a step forward from the previous education strategic plan.

The plan is not without weaknesses. In the evaluation literature, outcomes are typically thought of as measurable changes or absolute levels of performance that can be expected as a result of efforts to reach a goal. However, only three of the six environmental literacy outcomes in the education strategic plan and two of the three workforce development outcomes align with these expectations. The other outcomes describe strategies or processes that might contribute to reaching the goals. In addition, although diversity is a focus of the workforce goal, there is no mention of diversity or broadening participation in the environmental literacy goal. It is also unclear how NOAA can accomplish its goal of supporting the creation of a "world-class" workforce without a clear understanding of its own and the nation's workforce needs in the relevant areas. Although the importance of partnerships is stressed in the plan, there is no specific guidance about how or with whom to partner to connect to the national STEM infrastructure and human capacity. Finally, the use of the term "NOAA science" in the strategic plan is confusing. It is unclear whether this term is meant to refer to the science conducted by NOAA scientists, the research or the results of research funded by NOAA, or any science conducted on topics related to NOAA's mission.

Recommendations Regarding Education Goals and Outcomes

Recommendation II.1: NOAA education programs should formally address broadening participation of underrepresented groups as an important outcome through all phases, from the initial stages of planning through implementation and evaluation. The environmental literacy goal, in particular, should include outcomes related to reaching out to underserved and underrepresented communities.

Recommendation II.2: To reach NOAA's environmental literacy goal, the Education Council should develop its implementation plan and future revisions of the education strategic plan to:

- clarify how it will capitalize on scientific findings, engineering advances, and stewardship activities that relate broad national priorities to local concerns to engage individuals of all ages in education;
- articulate how NOAA education programs will draw on the scientific, engineering, research, and other expertise accessible within the agency as well as in the broader community;

- address the mismatch between the lack of an outcome related to stewardship and the focus on stewardship outcomes in local programs;
- consistently define outcomes as measurable concepts that allow an assessment of whether a goal is being reached, to clearly distinguish outcomes on audiences (impact) from outputs of activities; and
- provide more opportunities for local and regional education staff from all education programs to share effective practices and lessons learned.

Recommendation II.3: To achieve the workforce development goal, the education strategic plan, the education implementation plan, or both should call for periodic assessment of the current and anticipated needs in fields critical to NOAA's mission to guide investment in appropriate workforce development activities.

Recommendation II.4: NOAA education programs should draw from current and relevant scientific and engineering advances regardless of what agency, institution, or organization they are originated or funded by.

COMPOSITION AND MANAGEMENT OF THE EDUCATION PORTFOLIO

NOAA supports a wide range of education programs for varied audiences that include K-12, postsecondary, graduate, and informal education activities with local, regional, national, and international scope. NOAA has developed professional development programs, classroom materials, curricula, museum exhibits, place-based learning experiences, literacy documents, and other products. The audiences of the agency's education programs include teachers, students, scientists, and the public. A coherent, coordinated education portfolio is needed for achieving goals effectively and efficiently, for sharing successful strategies to engage and teach different audiences, for the pooling of resources to support synergistic activities, for developing cross-discipline activities, and for sustaining consistent education strategies.

Management of a federal education portfolio is complicated, and NOAA has characteristics that make it particularly challenging. Individual education programs may have separate mandates and often have local components with local control. Education programs are managed differently across the line offices and the Office of Education as a result of available resources for education (staff and funding), separate missions, and individual education mandates. The differences in management structures, missions, and education mandates are obstacles to creating a cohesive and coordinated education portfolio.

The Education Council is the primary means for NOAA to manage its education portfolio. Although relatively new, the Education Council, led by the Office of Education, serves an essential, high-level internal coordinating function. The Education Council led the development of the education strategic plan and is developing the collaborative working relationships necessary to implement it. However, the Education Council does not have budgetary or institutional control over the education efforts of the line and program offices, which limits its effectiveness in carrying out the agency's education mandate.

Since the education portfolio has developed in the absence of an overarching strategic direction and without a system to monitor or catalogue activities, it is difficult to assess its composition, balance, and impact. Such a system is needed to make informed decisions about the balance of the portfolio. What is clear, though, is that to date NOAA's education programs have been focused more on ocean or coastal concepts and issues and less on climate and atmospheric ones. Efforts are emerging to bring greater attention to climate and atmospheric issues and concepts across the agency's education activities.

Recommendation Regarding Composition and Management of the Education Portfolio

Recommendation III.1: NOAA should develop and implement a system to monitor and catalogue its education portfolio and guide decisions regarding what programs should be developed, continued, modified, or ended. In balancing the portfolio, the Education Council should

- increase attention to climate and atmospheric science education programs to complement the current focus on ocean science. These programs should emphasize the strong connections and interactions among the ocean, the atmosphere, the land, and human and nonhuman species;
- provide purposeful attention to both STEM learning and stewardship goals so as to enable synergies; and
- make decisions based on national education needs, the education priorities of the agency, and a clear picture of its education portfolio.

EDUCATION EVALUATION PRACTICES

The challenges of carrying out appropriate evaluations of education initiatives are large. Most federal science agencies are struggling to meet these challenges. NOAA is giving increasing attention to evaluating its education initiatives. The strategic education plan 2009-2029 highlights the need for more comprehensive evaluation of NOAA education initiatives.

NOAA has conducted evaluations of a small number of its educational activities, and these evaluations are limited in scope and quality. Summative evaluations have been carried out on a very small proportion of education activities, and there has been little consideration of evaluation that would enable it to recalibrate the entire portfolio to effectively meet its goals. The evaluations that have been carried out tend to focus on short-term and intermediate rather than long-term outcomes; rely on participant opinion, feedback, beliefs, and knowledge; and usually do not address outcomes related to attitudes or behavior. Outcome-based evaluations generally lack control or comparison groups or other ways to attribute potential changes solely to the education efforts themselves.

The Education Council is increasing its emphasis on evaluation and moving toward comprehensive program evaluation through the adoption of the Bennett Targeting Outcomes of Programs (TOP) model. The adoption of a uniform model as a framework to guide evaluation strategies and practices across all education programs is a useful step and may help implement a more strategic, coherent approach. However, as is the case with most evaluation models, the TOP model does not include specific guidance regarding the implementation of evaluations or how to design high-quality ones.

Data are needed for several purposes, including project monitoring, fiscal due diligence, and program evaluation. NOAA needs a systematic way of collecting data for each purpose to ensure that data are comparable across programs and initiatives and provide useful information for balancing the portfolio and assessing the strategic alignment of programs in it to the agency's overall goals.

Recommendations Regarding Education Evaluation Practices

Recommendation IV.1: The Education Council should continue to improve the evaluation expertise of its education program managers, contract with external evaluators for summative evaluation, and require the incorporation of the most appropriate and rigorous evaluation strategies during program development to guide design, continual improvement, and delivery of its education programs.

Recommendation IV.2: The Education Council should increase the emphasis on high-quality evaluations. Summative evaluations should focus on the program outcomes related to learning and stewardship, not only satisfaction with education experiences, and should use the most appropriate and rigorous evaluation designs.

Recommendation IV.3: The Education Council should consider developing a number of approaches to inform strategic portfolio management and how evaluation findings can be used to inform decisions about portfolio balance.

Recommendation IV.4: Education programs should evaluate internal collaboration among line offices and between education and operational and scientific staff, as well as the quality of external partnerships with other agencies, institutions, organizations, and the broader STEM communities.

EVIDENCE OF IMPACT

Although NOAA has created a large number of education initiatives with its limited education budget, there is evidence of impact for only a small proportion of them. The majority of initiatives that have collected information assessed only their scope and reach, which are not sufficient to judge impact. On the basis of the available evidence, all that can be said about the impact of such programs is that they are positively perceived by the participants. However, as yet, NOAA education programs serve a relatively small proportion of the nation's population.

There is a growing body of literature regarding effective practices in formal and informal science education, behavior change, reaching underserved populations, and workforce preparation. This literature can be used to support the development of science education programs that are likely to be successful.

Recommendation Regarding Evidence of Impact

Recommendation V.1: NOAA education staff should draw on evidence from education research, evaluations of NOAA programs, and external education expertise to identify and implement effective practices for supporting education activities.

Overall, NOAA's education staff is dedicated and passionate about addressing areas related to the agency's mission. Among NOAA's most valuable assets, they have developed diverse education activities for a wide range of audiences and regions. The agency is to be commended for its historic commitment to education, which precedes the agencywide congressional mandate on education. The agency's current education strategic plan is a significant improvement over the previous one. We hope that our recommendations continue to help NOAA improve its education efforts.

1

Overview and Introduction

The National Oceanic and Atmospheric Administration (NOAA) has long been a global leader in studying and communicating how the Earth's atmosphere and water systems influence people's lives and how they influence these systems. Education has been a component of NOAA's mission since it was created in 1970, with education projects covering a range of topics related to the agency's scientific and stewardship mission, including oceanic, atmospheric, climate, and environmental science. Many federal agencies, including NOAA, have resources that can improve the nation's understanding and interest in the relevant sciences and care of the planet. The importance of these interconnected fields and environmental stewardship cannot be overstated.

Although NOAA offices have long had mandates to engage in education activities, it was not until 2007 that it received an agencywide mandate for education. In that year, Congress, through the America Creating Opportunities to Meaningfully Promote Excellence in Technology, Education, and Science (COMPETES) Act (P.L. 110-69), explicitly directed NOAA to engage in education activities. The act calls for NOAA to "conduct, develop, support, promote, and coordinate formal and informal educational activities at all levels to enhance public awareness and understanding of ocean, coastal, Great Lakes, and atmospheric science and stewardship." In addition, the act called for NOAA to develop a 20-year education plan, to be reevaluated and updated every 5 years.

COMMITTEE CHARGE AND APPROACH

In recognition of the opportunity and obligation afforded by the new mandate, the NOAA Office of Education requested that the National Research Council (NRC) conduct a review of its existing education portfolio and plans for moving forward strategically. In response, the Board on Science Education of the NRC established the Committee for the Review of the NOAA Education Program to take stock of the existing education portfolio and review the education plan mandated by the America COMPETES Act. The committee included 14 members with expertise in the history and structure of NOAA education programs; education program evaluation; science and mathematics instruction at both the K-12 and higher education levels, with particular knowledge of oceanic, atmospheric, and climate sciences; teacher professional development; informal science education; workforce issues in the fields critical to NOAA's mission; and national, state, and local education policies and practices. Particular emphasis was given to including individuals on the committee with a working knowledge of diversity issues (see Appendix D for biographical sketches).

The committee's charge was to assess the agency's role in education, its educational goals and objectives, the impact of its education programs, the composition of its education portfolio, and how the agency conducts evaluations of its education programs. To address the charge, the committee reviewed NOAA's formal (K-12 and higher education) and informal education activities with the goal of helping the agency transform its various independent education programs and activities into a more coherent education portfolio aligned with common goals and outcomes. The committee assessed the agency's education initiatives in the context of its *Education Strategic Plan: 2009-2029* and evaluated its role in education, its program goals and objectives, its impact, its portfolio balance and priorities, and its evaluation approach. Communication and extension activities were explicitly omitted at NOAA's request because they are the focus of a separate report, *Engaging NOAA's Constituents: A Report from the NOAA Science Advisory Board* (2008; see Box 1.1 for that report's conclusions). In carrying out the study, the committee was cognizant that the boundaries between formal education, informal education, communication, outreach, and extension are porous and blurred. NOAA categorizes activities in each of these areas as education. For the purposes of this review, we adopted NOAA's definitions of these terms (see Box 1.2).

The committee followed an iterative process of gathering information, analyzing and deliberating on it, identifying gaps and questions, gathering additional information to fill these gaps, and carrying out further analysis. The limited time and resources for the study constrained the scope of the committee's review to existing documentation, site visits, and discussions

BOX 1.1
Communication and Extension Activities

In *Engaging NOAA's Constituents: A Report from the NOAA Science Advisory Board*, the NOAA Science Advisory Board called for the agency to dramatically change its approach to extension, outreach, and education in order to create true engagement with the public. The report has eight major conclusions:

1. A strategy for public engagement is missing.
2. There is no coordinating body to implement public engagement strategy.
3. There are insufficient resources for engagement.
4. Organizational culture in NOAA is not conducive to engagement.
5. The public is not fully aware of NOAA and its services.
6. NOAA is developing a new regional structure, although its place within the existing regional structure is not clear.
7. NOAA should better utilize partnerships in engagement.
8. NOAA should institutionalize a public accountability system.

The Extension, Outreach and Education Working Group recommended several steps that NOAA could take to improve its engagement strategy. These include devoting more resources to outreach and education, integrating outreach and extension into the Education Council's jurisdiction, and changing incentives so that employees are evaluated on the basis of their contributions to education and outreach programs. The report also recommended greater coordination and cohesion across NOAA's various offices and programs, with a focus on creating a more cohesive image for the public.

SOURCE: NOAA Science Advisory Board (2008).

with NOAA program and project staff. Although we did not carry out extensive original quantitative data collection, we did conduct two two-day site visits.

The committee held four public fact-finding meetings (see Appendix A for the meeting agendas). In addition, we reviewed documents related to NOAA's education portfolio, such as budget requests, project evaluations, project plans, and other technical reports.

During the first of the four public meetings, the committee heard presentations from and engaged in discussions with staff of the NOAA Office of Education, directors of education programs, and staff who oversee the education efforts of interagency ocean groups. At the second meeting, in addition to presentations about NOAA education projects, the committee heard presentations from and engaged in discussion with members of

BOX 1.2
NOAA's Definitions of Education, Outreach, Communication, and Extension

Formal Education: Learning within a structured education system in which children or adults are required to demonstrate proficiency.
Informal Education: Learning outside the established formal education system that meets clearly defined objectives through organized education activities.
Outreach for Education: Activities that are designed to build awareness, develop relationships, promote education products, and inspire educators, students, and the public to pursue further learning opportunities.
Communication: The process of delivering a message or other information through various media, whether verbal or nonverbal. The Office of Communications provides information about NOAA and the products and services it provides to the media, government officials, and the public.
Extension: Sustained interaction with specific audiences using education techniques to transfer science-based information or skills that inform decision making and/or change behavior.

SOURCE: National Oceanic and Atmospheric Administration (2009, p. 37).

interagency climate and atmospheric science and science education groups, and other marine, earth science, and environmental education organizations. The final two meetings included presentations from NOAA education programs, a discussion with a panel of experts on issues related to broadening participation in fields critical to NOAA's mission, and a set of presentations on NOAA's evaluation efforts in education.

The committee also had four background papers prepared.[1] One paper examined the existing external evaluations of NOAA's education projects (Brackett, 2009). A second paper summarized what is known about teaching and learning essential concepts in subjects related to NOAA's mission (Tran, 2009). A third paper described issues related to participation of underserved and underrepresented populations in fields related to NOAA's mission (Levine et al., 2009). A fourth paper explored how well the multiple, often small education programs spread across NOAA serve the larger educational goals of the agency (Clune, 2009). As noted, committee members conducted two site visits (see Appendix B for site visit agendas) to interview education specialists, experience NOAA-sponsored activities firsthand, and learn how education programs are managed.

---

[1]See http://www7.nationalacademies.org/bose/NOAA_Commissioned_Papers.html [accessed May 2010].

NATIONAL EDUCATION NEEDS RELATED TO NOAA'S MISSION

In order to review NOAA's role in education, it was necessary to identify and understand the need for education in areas related to the agency's mission, as well as the various ways that other federal agencies are or could be involved in education.

Meeting the many needs of the nation requires a diverse workforce, adequate in number, with expertise across a broad range of scientific and technical disciplines. This workforce should understand how to apply science, technology, engineering, and mathematics (STEM) to the solution of problems confronted every day throughout society. A democratic society needs all citizens to be scientifically literate in order to participate fully in national debates about urgent policy issues, such as climate change and alternative fuels. The importance of NOAA's role in education is directly related to the need for a scientifically literate society and a well-prepared workforce in fields related to its mission.

As far back as 1929, the NRC emphasized that the United States would realize advances in science and mathematics knowledge only by training a workforce that is knowledgeable about science, is sufficient in size and ability, and is afforded continual educational opportunities (National Academy of Sciences, 1929). In the past five years there has been increased attention to STEM and environmental education issues. The America COMPETES Act and some high-profile reports highlight the critical importance of science education to the future of the world economy and public literacy. The National Academies report, *Rising Above the Gathering Storm* (National Academy of Sciences, National Academy of Engineering, and Institute of Medicine, 2007), pointed out the declining competitiveness of the United States in terms of scientific competence and discoveries and made recommendations to counteract this downward spiral. Paramount is an emphasis on the high-quality education necessary to produce the next generation of scientists and engineers.

Currently, it is not clear that a sufficient science and engineering workforce is being prepared. The Congressional Research Service (2008) reports that, although the number of degrees in some STEM fields (particularly biology and computer science) has increased, the overall proportion of STEM degrees awarded in the United States has historically remained at about the same low percentage. In 2002-2003, STEM degrees were about 15 percent of all associate degrees, 17 percent of baccalaureate degrees, 13 percent of master's degrees, and 35 percent of doctoral degrees. The United States ranks about 20th among nations in the proportion of 24-year-olds who earn degrees in natural science or engineering. In addition, enrollment of U.S. citizens in graduate science and engineering programs has lagged that of foreign students in these programs. According to the National Science Foundation (NSF) Survey of Earned Doctorates, foreign students earned one-third of all doctoral degrees awarded in the United States in 2003. Still, the silver lining here is that between 50 and 60 percent of these foreign students become integrated into the U.S. workforce, and many remain in the United States for years, if not permanently.

While precise estimates have not been developed, there is a growing concern that in the next 10 years perhaps as many as half of the skilled workforce presently occupying critical positions in the federal government will retire (U.S. Commission on Ocean Policy, 2004). In addition, the mix of skills needed to bring the ocean, atmosphere, and climate workforce into the new millennium is not clear (U.S. Commission on Ocean Policy, 2004). Overall, the United States risks a competitive disadvantage in the global marketplace and increasing environmental problems. The disparity of individuals from ethnic minority or underrepresented populations in these scientific fields is a particular concern (Cucker, 2001; Huntoon and Lane, 2007; Levine, González, and Martínez-Sussmann, 2009).

In today's competitive world of knowledge-based and technology-driven economies, the global market (including living resources, transportation, urbanization, tourism, and renewable and nonrenewable energy) and associated services (engineering, transportation, environmental monitoring, resource management, policy and governance, education, etc.) depend on increasing knowledge concerning the atmosphere and land and marine environments. This requires that the federal government, states, municipalities, and the public make informed decisions. Moving forward also necessitates a capable and informed workforce.

The picture of science education illuminated by recent reports (e.g., Congressional Research Service, 2008; National Research Council, 2006; U.S. Commission on Ocean Policy, 2004; U.S. Department of Education, 2007) also illustrates many shortcomings related to the development of a scientifically literate society. Relatively few students relate to scientific concepts or understand scientific processes (American Association for the Advancement of Science, 1989), and the nation does not perform as well as other developed countries in science and mathematics education (Lemke and Gonzales, 2006; Mullis et. al., 2007). Likewise, literacy issues have been noted in areas specifically related to NOAA's mission. The American public has only a superficial awareness of the importance of the ocean, the atmosphere, and climate in daily life, let alone its importance to all life on the planet (Belden et al., 2001; U.S. Commission on Ocean Policy, 2004). There are increasing calls for improved oceanic, atmospheric, and climate education, yet there is evidence that the amount of time devoted to science education in the elementary grades of some states has decreased since the

No Child Left Behind (NCLB) legislation became a law (Center on Education Policy, 2007; Dorph et al., 2007).

Clearly there is a need to address the shortcomings in science education, particularly in areas related to NOAA's mission. Federal agencies have an important role and have long been important actors in STEM education. Although not all federal science agencies have an explicit education mission, most have made efforts to reach out to students, teachers, and the public to inform them about STEM issues. Agencies that have developed education programs related to planetary, environmental, and scientific processes include NOAA, the U.S. Navy, the National Aeronautics and Space Administration, NSF, the U.S. Environmental Protection Agency (EPA), the Minerals Management Service, the U.S. National Park Service, the U.S. Fish and Wildlife Service, and the U.S. Geological Survey.

A persistent concern about these federal education programs is that they are not well coordinated across or even within agencies, and funding mechanisms and resources are seldom well utilized (Association of American Colleges and Universities, 2006; Business Roundtable, 2005; Federal Coordinating Council for Science, Engineering and Technology, 1992; Lewis, 2005; National Research Council, 2006; National Summit on Competitiveness, 2005; U.S. Commission on Ocean Policy, 2004; U.S. Department of Education, 2007). Even within individual agencies, offices that carry out education initiatives often do not collaborate or communicate with each other.

The lack of coordination across federal agencies, combined with the increasing urgency of developing the nation's workforce and scientific literacy needs, heightens the need to carefully consider the potential role of NOAA and other agencies in the national education landscape. The history and structure of the agency provide an important context for understanding NOAA's role.

HISTORY AND STRUCTURE

In a speech before Congress in July 1970, President Richard Nixon created NOAA by executive order by combining a number of federal agencies. Its creation, coupled with the creation of the EPA, was part of an effort to unify the nation's widely scattered, piecemeal environmental activities and provide a rational and systematic approach to understanding, protecting, developing, and enhancing the environment. NOAA was to lead the development of a consolidated national oceanic and atmospheric research and development program and provide a variety of scientific and technical services to other federal agencies, private-sector interests, and the public. Specifically, NOAA was established as a science-based agency, responsible for predicting changes in the oceanic and atmospheric environments and living marine resources and providing related data, information, and services.

Three of the agencies initially brought together under NOAA were the oldest agencies in the United States: the U.S. Coast and Geodetic Survey, the U.S. Environmental Science Services Administration (ESSA), and the Bureau of Commercial Fisheries. The U.S. Coast and Geodetic Survey was renamed the National Geodetic Survey, housed in NOAA's National Ocean Service. The mission of ESSA, which had been created in 1965 as part of the U.S. Department of Commerce, was to oversee the nation's weather and climate operations and included the Weather Bureau and National Data Center. In January 1966, ESSA changed the Weather Bureau's name to the National Weather Service (NWS), and the National Data Center was renamed the Environmental Data Service (EDS). The U.S. Commission of Fish and Fisheries—the nation's first federal conservation agency, initiated in 1871 to protect, study, manage, and restore fish—became NOAA's National Marine Fisheries Service, or NOAA Fisheries. Since its creation in 1970, NOAA has continued to expand through the incorporation of other agencies and the creation of new programs. The agency currently has six line offices:

1. Office of Oceanic and Atmospheric Research,
2. National Ocean Service,
3. National Environmental Satellite Data and Information Services,
4. National Marine Fisheries Service,
5. National Weather Service, and
6. Program Planning and Integration.

These were incorporated into the agency at different times and through different processes (see Box 1.3 for details on the six line offices; see Figure 1.1 for the 2008 organizational chart). The focus of each line office is determined by its distinct missions and congressional mandates.

NOAA's mission statement does not include education, yet dedication to education is underscored by the agency's vision to develop a society that is informed on issues related to its mission (National Oceanic and Atmospheric Administration, 2009). Also, there are a number of education objectives within each of the agency's four broad goals: (1) to protect, restore, and manage coastal and ocean resources; (2) to understand climate variability and change to enhance society's ability to plan and respond; (3) to serve society's needs for weather and water information; and (4) to support commerce with information for safe, efficient, and environmentally sound transportation (National Oceanic and Atmospheric Administration, 2009).

BOX 1.3
NOAA Line Offices

The NOAA website provides these descriptions of the six line offices:

Office of Oceanic and Atmospheric Research (OAR – http://www.oarhq.noaa.gov) provides the research foundation for understanding the complex systems that support our planet. Working in partnership with NOAA's other organizational units, OAR makes better forecasts, earlier warnings for natural disasters, and a greater understanding of the Earth possible. OAR's role is to provide unbiased science to better manage the environment, nationally and globally.

National Ocean Service (NOS – http://www.nos.noaa.gov) keeps ocean and coastal areas safe, healthy, and productive. The NOS serves America by conserving marine and coastal places for present and future generations, ensuring safe and efficient maritime transportation, and promoting innovative science and technology solutions to coastal challenges.

National Weather Service (NWS – http://www.weather.gov) provides weather, hydrologic, and climate forecasts and warnings for the United States, its territories, adjacent waters, and ocean areas for the protection of life and property and the enhancement of the national economy. NWS data and products form a national information database and infrastructure for use by other government agencies, the private sector, the public, and the global community.

National Marine Fisheries Service (NMFS – http://www.nmfs.noaa.gov) is responsible for the management, conservation, and protection of living marine resources within the United States' Exclusive Economic Zone. NMFS assesses and predicts the status of fish stocks, ensures compliance with fisheries regulations, works to reduce wasteful fishing practices, and recovers protected marine species without unnecessarily impeding economic and recreational opportunities.

National Environmental Satellite, Data, and Information Service (NESDIS – http://www.nesdis.noaa.gov) provides timely access to global environmental data from satellites and other sources to promote, protect, and enhance the Nation's economy, security, environment, and quality of life. To fulfill its responsibilities, NESDIS acquires and manages the Nation's operational environmental satellites, provides data and information services and conducts related research.

Office of Program Planning and Integration (PPI – http://www.ppi.noaa.gov) provides corporate management to coordinate NOAA's many lines of service with the nation's many needs for environmental information and stewardship. It ensures that investments and actions are guided by a strategic plan; are based on sound social and economic analysis; adhere to executive and legislative science, technology, and environmental policy; and integrate the full breadth of NOAA's resources, knowledge, and talent to achieve its mission.

SOURCE: See http://www.dco.noaa.gov/transition/structure/lineoffices.html [accessed May 2010].

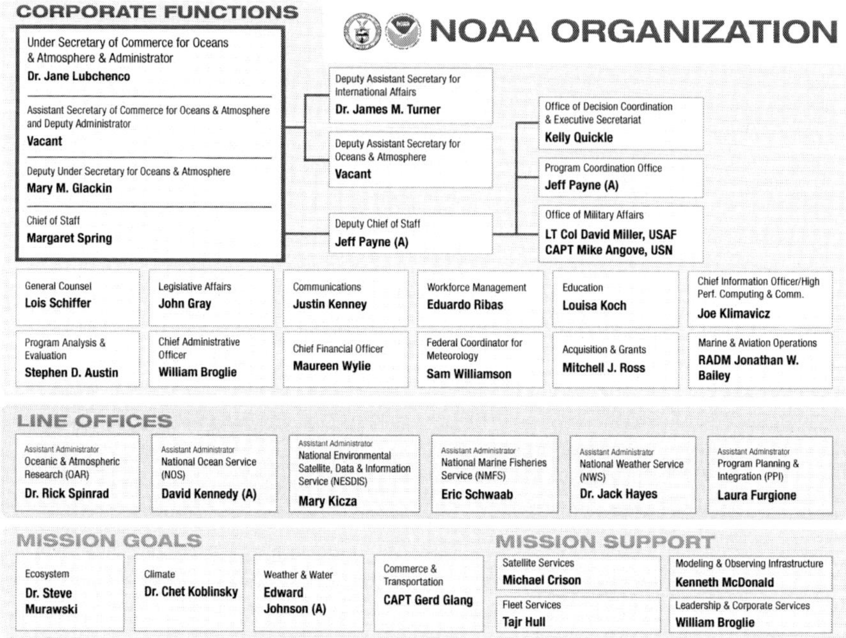

FIGURE 1.1 NOAA 2008 organization chart.
SOURCE: National Oceanic and Atmospheric Administration.

Education at NOAA

When NOAA was created, its operational mission was well defined, but without any agencywide role in education. Since its creation, individual programs have received mandates to be involved in education. For example, in 1970 NSF's National Sea Grant Program, an academic/industry/government partnership to enhance the nation's education, economy, and environment, became part of NOAA and was mandated to support education activities. Two years later, the National Marine Sanctuaries Program, which is also mandated by law to include education initiatives, was established by the Marine Protection, Research, and Sanctuaries Act of 1972 (P.L. 92-532). Also in 1972, the National Estuarine Research Reserve System, which has a mandate to conduct education activities, became part of NOAA. Box 1.4 lists the other programs and the laws that mandate them to support education throughout NOAA's history.

As noted, the America COMPETES Act recently defined NOAA's educational role: to advance environmental literacy; to promote a diverse workforce in ocean, coastal, Great Lakes, weather, and climate sciences; and to encourage stewardship and informed decision making for the nation. NOAA fulfills this role through its Office of Education and its Education Council. The Education Council, which includes leadership from all of NOAA's major education initiatives, was created to coordinate education activities across NOAA and oversee the development and implementation of its strategic education plan.

Brief Education Program Descriptions

Five of NOAA's line offices—the National Ocean Service, the Office of Oceanic and Atmospheric Research, NWS, the National Marine Fisheries Service, and the National Environmental Satellite, Data, and Information Service—along with the Office of Education, develop and support the majority of the agency's education activities. In addition, NOAA provides an unquantified number of scientific research grants and contracts that include education (via support of student research and other efforts) as part of their research program. The committee focused its review on the more robust education programs across the five line offices of the agency and the Office of Education. Brief descriptions of these programs are provided below, and more detailed descriptions can be found in Chapter 4. The range of activities across these programs illustrates the breadth of NOAA's education activities and the resulting challenge inherent in developing a coordinated, agencywide approach to education.

Overall, NOAA is supporting education projects that address the needs of individuals of varied ages and backgrounds, and the projects reflect the interdisciplinary nature of the scientific research, stewardship activities, technology development, and engineering design of the agency. The educational portfolio includes activities that occur in formal and informal learning environments. Across these learning environments, NOAA primarily provides science education and environmental science education activities, which overlap but have distinct features.

The majority of science and science education in NOAA is categorized as earth systems science: the study of the unified set of physical, chemical, biological, and social components, processes, and interactions that together determine the state and dynamics of Planet Earth, with an emphasis on observing, understanding, and predicting global environmental changes (Earth System Science Partnership, 2009). Through its stewardship role, NOAA also engages in environmental science and environmental education activities.

BOX 1.4
Education Mandates

The America COMPETES Act (P.L. 110-69) requires that NOAA create a 20-year education strategic plan, and that "the Administrator of the National Oceanic and Atmospheric Administration shall conduct, develop, support, promote, and coordinate formal and informal educational activities at all levels to enhance public awareness and understanding of ocean, coastal, Great Lakes, and atmospheric science and stewardship by the general public and other coastal stakeholders, including underrepresented groups in ocean and atmospheric science and policy careers. In conducting those activities, the Administrator shall build upon the educational programs and activities of the agency" (Sec. 4002). In addition to this charge, NOAA is to be a full participant in any interagency effort promoting innovation and/or economic competitiveness.

The Coral Reef Conservation Act (P.L. 106-562) requires that any activities funded by this act also enhance public awareness, education, understanding, and appreciation of coral reef ecosystems.

The Coastal Zone Management Act (P.L. 109-58); Section 1461 discusses the National Estuarine Research Reserve System and requires NOAA to acknowledge that any designation of a reserve provides the opportunity to enhance public awareness and understanding of estuarine areas and provides suitable opportunities for public education and interpretation.

The Magnuson-Stevens Fishery Conservation and Management Act (P.L. 109-479) contains some very specific restrictions on NOAA education. For example, before entering a Pacific Insular Area fishery agreement, the marine conservation plan should include (but not be limited to) conservation, education, and enforcement activities related to marine and coast management. It also requires that there should be grants to the University of Hawaii for technical assistance projects by the Pacific Island Network, such as education and training in the development and implementation of sustainable marine resources development projects, scientific research, and conservation strategies. Other objectives in-

National Ocean Service

The National Ocean Service runs three major education programs, and a suite of education projects and materials are managed by the National Ocean Service corporate office.

The *National Estuarine Research Reserve System* (NERRS) is a network of 27 protected areas established for long-term research, education, and stewardship. It is a partnership program between NOAA and the coastal states that protects more than 1 million acres of estuarine land and water. Educational programming linked to research and stewardship has been incorporated at the reserves since their inception in 1972. The goals of NERRS education activities are to enhance public awareness and under-

clude the establishment of a pilot program for regionally based marine education and training programs in the Western Pacific and the Northern Pacific that focus on stewardship of living marine resources. This would include the establishment of "programs or projects that will improve communication, education, and training on marine resource issues throughout the region and increase scientific education for marine related professions among coastal community residents, including indigenous Pacific islanders, Native Hawaiians, Alaskan Natives, and other underrepresented groups in the region." There are additional educational objectives targeting Western Pacific Demonstration Projects and an important study on the shortage of individuals with postbaccalaureate degrees in subjects related to fishery science. The secretaries of commerce and education were tasked to transmit a report to Congress detailing the findings and recommendations of the study under this section.

The National Marine Sanctuary Program supports research, monitoring, evaluation, and education programs that are in line with the National Marine Sanctuaries Act (P.L. 106-513, Sections 1431 et seq.). Under this act, educational efforts must emphasize the conservation goals and sustainable public uses of national marine sanctuaries. This would include any of the target audiences (the public, teachers, students, national marine sanctuary users, and ocean and coastal resource managers) as well as any interpretive facilities that are constructed or developed.

The National Sea Grant College Program Act (33 U.S.C. 1123(d)(3)(B)) has been amended by the addition of the following phrase: "encourage and promote coordination and cooperation between the research, education, and outreach programs of the Administration and those of academic institutions," which clearly supports the enhanced coordination between and within agencies. This is also spelled out in Section 9 of the National Sea Grant College Program Act Amendments of 2002 (H.R. 3389), which requires an annual report on how NOAA will accomplish this goal.

standing of estuarine areas and provide suitable opportunities for public education and interpretation. Most reserves also provide K-12 education, ranging from hands-on field experiences for students to professional development opportunities for teachers.

The 13 *National Marine Sanctuaries*, established by the Marine Protection, Research, and Sanctuaries Act of 1972, and a National Marine Monument, established by the Antiquities Act of 2006, promote public understanding of national marine heritage and the marine environment. Educational materials for students and teachers are provided online through the Office of National Marine Sanctuaries, and hands-on education experiences are also available at each sanctuary. Oceans Live is a national program that connects the Office of National Marine Sanctuaries to the public through telepresence. It links existing systemwide oceanographic monitoring programs with interactive telepresence technology.

The *Coral Reef Conservation Program* (in multiple offices) was established in 2000 and is a partnership between the NOAA line offices working on coral reef issues, including the National Ocean Service, the National Marine Fisheries Service, the Office of Oceanic and Atmospheric Research, and the National Environmental Satellites, Data, and Information Service. The program brings together expertise from NOAA's line offices for a multidisciplinary approach to managing and understanding coral reef ecosystems. The education activities are focused on the three major threats of climate change, land-based sources of pollution, and fishing impacts. The goal of these education activities is to promote an informed society that understands the value of coral reef ecosystems, the threats they face, and the actions individuals can take to reduce human impacts on them.

The National Ocean Service corporate office education team is responsible for the development of teacher professional development opportunities, online curricular resources, and partnerships with groups such as the National Science Teachers Association and the Council of State Science Supervisors.

Office of Oceanic and Atmospheric Research

The Office of Oceanic and Atmospheric Research (OAR) line office runs three major education programs.

The *National Sea Grant Program* (Sea Grant) is a nationwide network of university-based programs that aims to foster environmental stewardship, long-term economic development, and responsible use of the nation's coastal, oceanic, and Great Lakes resources. This program was established almost 40 years ago and consists of 30 individual programs based at universities across the country, the Sea Grant Educators Network, the Dean John A. Knauss Marine Policy Fellowship, and the Graduate Fellowship Program. The program has a long tradition of supporting environmental literacy through education, working with K-12 teachers, bringing students out of the classroom and into the natural environment, and supporting undergraduate and graduate students.

The *Ocean Explorer Program* (a partnership between the National Ocean Service and the Office of Oceanic and Atmospheric Research), serves as a public archive of the exploration program, chronicling many of the missions with daily logs, background essays, and multimedia offerings. Educational materials are developed through collaborations between ocean explorers and teachers. In addition, the recent addition of the NOAA ship *Okeanos Explorer*, which travels around the globe to map the seafloor and characterize largely unknown areas of the ocean, supports education activities through use of real-time broadband satellite communications to connect the ship and its discoveries with audiences ashore.

The *Climate Program Office* takes an audience-focused approach to promoting climate science literacy among the public. The program office communicates the challenges, processes, and results of NOAA-supported climate science through stories and data visualizations on the web and in popular media. It provides information to a range of audiences to enhance society's ability to plan and respond to climate variability and change. The office led the interagency development of climate literacy principles and is developing a variety of electronic and professional development programs.

Office of Education

The Office of Education also runs three major educational programs and two scholarship initiatives.

Environmental Literacy Grants provide funding for environmental literacy projects in support of K-12 and informal education. Funded projects are between one and five years in duration and promote changes in K-12 or informal education to expand the amount of earth systems science taught in the classroom and improve student learning and application of that subject. Projects are reviewed based on incorporation of NOAA data, data visualizations, and resources and are encouraged to further the use of earth systems science concepts related to NOAA's mission goals, such as the concepts articulated in the Ocean Literacy and Climate Literacy Essential Principles and Fundamental Concepts.

The *Bay-Watershed Education and Training Program* provides grants in support of locally relevant experiential learning in the K-12 system. The program currently operates in the Chesapeake Bay area, California, and Hawaii and just recently expanded to include three new regions: the Northeast, the Gulf of Mexico, and the Pacific Northwest. Funded projects involve watershed educational experiences addressing regional priorities and provide hands-on watershed education to students and teachers to foster stewardship. The program also claims to support larger scale impacts, such as systematic, long-term professional development for teachers to improve their capacity to teach, inspire, and lead young people toward thoughtful stewardship.

The *Educational Partnership Program* provides financial assistance through four competitive program components: the Cooperative Science Centers, the Environmental Entrepreneurship Program, the Graduate Sciences Program, and the Undergraduate Scholarship Program. The program focuses on underrepresented populations in fields related to NOAA's mission through partnerships with minority-serving institutions. It consists of training initiatives designed to address the full spectrum of capacity-building opportunities, including student training, peer and collaborative research, and faculty staff exchanges.

The *Dr. Nancy Foster Scholarship Program* was authorized by Congress soon after her death in June 2000, as a means of honoring the marine biologist's work and contribution to the nation. The program recognizes outstanding scholarship and encourages independent graduate-level research, particularly by female and minority students, in oceanography, marine biology, and maritime archaeology.

The *Ernest F. Hollings Scholarship Program* was created in 2005 to honor retiring Senator Ernest F. Hollings. It provides undergraduate students with awards that include academic assistance; a 10-week, full-time internship position at a NOAA facility during the summer; and academic assistance for a second academic year if the student is reappointed. The purpose of the internship after the first year of the award is to provide scholars with hands-on educational training experience in NOAA-related science, research, technology, policy, management, and education activities.

National Weather Service

The NWS provides a variety of educational resources to classrooms and the public, although it does not have a concerted education program. StormReady and TsunamiReady are its major national programs with an education component. They are community preparedness programs that use a grassroots approach to helping communities develop plans to handle all types of severe weather (e.g., thunderstorms, tornados, hurricanes) and tsunamis. Information provided by StormReady and TsunamiReady includes targeted publications about severe weather safety for public and community leaders, related statistics, presentations, brochures, and community seminars to promote the importance of public readiness. Most of the remaining education activities are informal, such as the NOAA/American Meteorological Service WeatherFest, an interactive science festival for the public that includes a teacher training workshop. The NWS has supported formal education activities, such as the creation of the Xtreme Weather CD. In addition, approximately 2,400 school visits are made by staff of the field offices each year.

National Marine Fisheries Service

The National Marine Fisheries Service is responsible for stewardship of the nation's living marine resources and their habitats within the United States' Exclusive Economic Zone (waters 3 to 200 nautical miles offshore). Currently it has no education program as such, but rather a variety of education projects carried out by the six regional science centers. These projects are highly decentralized and therefore are largely determined by regional centers, and many of them are informal and combine outreach and education. The most prominent ones are Teacher at Sea (started in 1990) and Teacher in the Air (started in 2004). Management of these two programs was transferred to the National Marine Fisheries Service during our review. The programs place teachers on a research vessel or plane, an activity that allows kindergarten through college-level teachers to work under the tutelage of scientists and crew aboard NOAA research survey ships or aircraft.

National Environmental Satellite, Data, and Information Service

The National Environmental Satellite, Data, and Information Service provides the largest active archive of weather data in the country and includes the National Climatic Data Center, the National Oceanographic Data Center, and the National Coastal Data Development Center. The education efforts of this office are primarily electronic, including their collaborative efforts on the Coral Reef Conservation Program.

ORGANIZATION OF THE REPORT

This report provides guidance to NOAA's continued efforts to support environmental literacy of the nation and meet the workforce needs of the agency and the nation. As we conducted our review and prepared this report, five major guiding considerations emerged:

1. The importance of NOAA's coordinating role in federal education efforts related to earth system science and stewardship and attention to defining this role in partnership with other federal science agencies.
2. The challenge NOAA faces in addressing both its needs as a science agency and national education and workforce needs, which is heightened by the urgency of calls for improving public literacy in science so that all citizens can participate in public discussions of science-related issues.
3. The urgency of significantly increasing diversity in the audiences reached by NOAA's education programs and translating this into diversity of the workforce in NOAA and diversity in sciences and technology workforces in the nation in general.
4. The programmatic, management, and financial complexities inherent to developing a portfolio of education activities that balances the agency's science, environmental, education, and stewardship goals.

5. The need for an approach to the education portfolio that emphasizes continual improvement and supports rigorous evaluation of NOAA education programs.

These considerations are discussed across Chapters 2 to 5, and they influenced the conclusions and recommendations presented in Chapter 6.

Following this introduction, Chapter 2 provides an overview of the national education landscape in which NOAA programs operate and outlines how NOAA can define its role in this context. Brief descriptions of the education efforts by other agencies are provided, and the assets and limitation that define NOAA's role in education are described. The need for coordination and collaboration across federal and state agencies, institutes of higher education, K-12 education communities, and private-sector and nonprofit organizations is stressed.

Chapter 3 presents the NOAA education portfolio, critiques the education strategic plan, and outlines effective education practices for professional development for teachers, curriculum, instruction, informal education environments, promoting diversity, and addressing workforce needs. The strengths and weaknesses of the strategic education plan and effective education practices to support the agency's education plan are described.

Chapter 4 evaluates the major NOAA education projects based on briefings from NOAA staff, administrative documents, annual reports, recent external evaluations, and research in education regarding effective practices. Each of the major education programs is described, and the evidence of their impact is highlighted. Cross-cutting issues, including portfolio balance, are discussed at the end of the chapter.

Chapter 5 critiques the framework NOAA has selected to guide future project evaluations and the previous project evaluations. The chapter provides guidance on implementing the education evaluation framework that NOAA has selected.

Chapter 6 presents the committee's conclusions and recommendations, which specifically address the appropriate role of NOAA in education, the appropriate goals and outcomes of its education activities, effective and scalable evaluation strategies to assess the impact of education projects and the education portfolio, the appropriate balance of its education portfolio, and the impact of existing education activities.

2

NOAA's Role in the Education Landscape

This chapter describes the complex education landscape in which the National Oceanic and Atmospheric Administration (NOAA) operates, the challenges and opportunities it faces, and the factors it should consider in developing its role in education. We provide a general overview of science education in the United States and discuss why it is appropriate for federal agencies, including NOAA, to have a role in supporting oceanic, atmospheric, climate, and environmental education.

THE SCIENCE EDUCATION SYSTEM

NOAA is one of many agencies, institutions, and organizations working to improve the nation's science literacy and technical workforce. To understand its role in education it is critical to understand the major players in the system and the role of the federal government in the education system. This section outlines the roles of various entities in the K-12, higher, and informal education systems.

K-12 Education

The responsibility for public education is not specified in the U.S. Constitution; hence, individual states have the right and responsibility for K-12 public education. Schools, school administrators, and teachers are held accountable within their state system. Federal agencies and nonprofit and private-sector organizations can offer advice, materials, training, funding, and other support. In 2005-2006, about 90 percent of the approximately

$520 billion spent on K-12 education in the United States came from state and local governments, and about 10 percent came from the federal government (U.S. Department of Education, 2008).

The most important state responsibilities are (1) developing state education standards and associated support (e.g., curriculum guides, curriculum frameworks), (2) implementing statewide high-stakes assessments and enforcing consequences of high or low performance on the assessments, and (3) credentialing teachers and establishing criteria for teacher licensing. In addition, in about 40 percent of states, the state department of education reviews and approves curriculum materials on a statewide basis.

Standards determine the direction and nature of science learning and set learning goals for all students. Establishing standards is an important tool for states to influence science instruction. Establishing standards is a process in which the content of science learning is melded into a montage of learning expectations drawn from a variety of sources; primarily the *National Science Education Standards* (National Research Council, 1996), the *Benchmarks for Science Literacy* (American Association for the Advancement of Science, 1993), and local science teaching traditions. In addition, the way state standards are implemented may be influenced by informal education institutions and organizations in the state, including some that receive funding from federal sources, such as the National Science Foundation (NSF), the National Aeronautics and Space Administration (NASA), NOAA, and others.

The involvement of the federal government in K-12 science education dates back to the mid-20th century. The federal government currently influences the national agenda in K-12 science, technology, engineering, and mathematics (STEM) education through two processes. First, legislation affects federal funding, which can lead to changes in state and local education systems. For example, Title I of the Elementary and Secondary Education Act, and its reauthorization under the No Child Left Behind (NCLB; U.S. Department of Education, 2004) Act, implemented by the U.S. Department of Education, has significant effects on K-12 STEM education. NCLB requires states to establish academic standards for reading, mathematics, and science in order to apply for specific types of federal support. NCLB also requires states to hold schools and districts accountable for student performance in reading and mathematics. This has resulted in increased instructional time in reading and mathematics to make adequate yearly progress and consequently little time for science instruction (Griffith and Scharmann, 2008). This is particularly true in Title I schools (schools in which 40 percent or more of the students come from low-income families), in which failure to make adequate yearly progress carries significant consequences for the use of federal funding received by the school.

Second, Congress provides funding for federal agencies involved in K-12 STEM education, which influences the types of education programs that are developed and supported by federal agencies. Even though the influence of the federal government on education has grown, its authority over K-12 public education remains limited. The federal government does not set a national curriculum or mandate state or local participation in federal programs. States can refuse to participate in any federal education program, including NCLB (thereby forgoing its associated funds). Nevertheless, however small the amounts of funding might be, the opportunity to receive federal financial support can and does influence the direction of science education.

Many federal agencies, including the Department of Education, NSF, the Department of Health and Human Services, the Department of Energy, the Department of Commerce, the Department of Agriculture, and the Department of Transportation, NASA, and NOAA, fund K-12 STEM education programs and research. These agencies share their expertise in science and science education through their involvement in education programs for students and teachers at the K-12 level. They develop programs that provide opportunities for learners to understand the nature of science. Agencies with education programs that are deployed across many states (NOAA is one) influence the context and nature of instruction through professional development opportunities and instructional resources offered to teachers. For example, the influence of NOAA can be seen in the curricula of coastal states (Hoffman and Barstow, 2007).

Nonprofit and private-sector organizations also influence the K-12 education system through various avenues, including managing private schools, advocating for policy reform, funding education initiatives, contributing to the standards-making process, among others. And increasingly, they are calling for a higher level of societal awareness about critical issues that require a STEM education. Numerous reports focus directly or indirectly on STEM and environmental education, for example the Kauffman Foundation (2007), the William and Flora Hewlett Foundation (Plumb and Reis, 2007), and the Carnegie Corporation (Commission on Mathematics and Science Education, 2009). Private-sector organizations have highlighted the national need to close achievement gaps with other countries and raise achievement levels for all U.S. students to meet and exceed international benchmarks (McKinsey and Company, 2009; Swanson, 2009) and highlight the importance of building student knowledge and skills to meet the needs of the 21st century global labor market (National Center on Education and the Economy, 2006). They have called for common and rigorous standards, better assessments, and improved human capital management with a focus on recruiting, training, and retaining the best teachers (National Academy of Sciences, National Academy of Engineering, and Institute of Medicine, 2006).

Higher Education

In the United States, colleges and universities have the primary responsibility to provide STEM undergraduate and graduate education. Individual departments in colleges and universities have significant autonomy to design curriculum, award degrees, and construct educational experiences for their students. Departments are influenced by their own institution's policies (and for public universities and colleges, these policies are in turn influenced by the state); national higher education accreditation agencies, policies, and reports; private-sector organizations, including discipline-specific professional societies; and the requirements of agency employment.[1] Federal influences on STEM curriculum are in the form of direct scholarship and fellowships to students, indirect support for students through their participation in funded research, and support for curriculum development and education research in higher education.

The federal contribution to STEM higher education comes from its support of the university research that students engage in or by directly funding students themselves, through fellowships or scholarships. In 2005, the combined federal contribution to research at universities was nearly $16 billion, of which $9 billion was from the U.S. Department of Health and Human Services (National Science Foundation, 2005). NSF is the next largest contributor to research at universities and colleges; its annual budget represents 21 percent of the total federal budget for basic research conducted at America's colleges and universities. The U.S. Department of Commerce, the U.S. Department of Energy, the U.S. Department of Defense, and NASA, all fund extramural research programs that support STEM education as well, in smaller amounts. By awarding funding to research aligned with their mission, they catalyze training and instruction in certain areas, thereby indirectly influencing the curriculum that students experience. For example, the Educational Partnership Program invests in areas of increasing interest to NOAA and helps develop institutional capacity in those areas, including contributing to the creation of new degree programs.

Like the federal government, states also contribute to STEM education through scholarships and fellowships to students in STEM disciplines and support for STEM research in which students participate. However, state governments play a much smaller role in higher education than in K-12 education. States do not credential STEM professors at colleges or univer-

[1]For example, a meteorology degree in a public university is ultimately designed and delivered by the faculty of the university, but it may be constrained by university policies that define a minimum number of credits and set of required courses and state policies that define a maximum number of required credits per degree. In addition, faculty will consider degree recommendations from the American Meteorological Society and even the employment requirements described by the National Weather Service.

sities, define education standards, or implement any testing of university or college students. States do provide partial funding for their public institutions, although the amount varies considerably. States are the primary funders of state, junior, and some four-year teacher colleges; private institutions receive a much smaller proportion of their funding from the state. For institutions to which they provide a significant amount of funding, states can exercise fiscal oversight, influence admissions policies, and influence the curriculum in the broadest terms. In fiscal year (FY) 2008, state and local funding for public and independent higher education totaled $89.2 billion (State Higher Education Executive Officers, 2008). States exert very little influence on private institutions of higher education.

Private-sector organizations also play a role in the higher education system. Some professional societies influence higher education in STEM by setting professional standards associated with many degrees. By defining standards for admission to the society, many more indirectly influence the degree requirements of the associated field. Many societies also provide forums for improving discipline-related teaching practice in higher education, and offer scholarships to students. In addition, some organizations advocate for and fund initiatives to improve the higher education systems. For example, recent reports (Association of American Colleges and Universities, 2007; Building Engineering and Science Talent, 2004; National Academy of Engineering, 2005; National Academy of Sciences, National Academy of Engineering, and Institute of Medicine, 2006; National Science Board, 2003, 2004, 2009; Project Kaleidoscope, 2006) have outlined the importance of higher education in addressing the crises in U.S. STEM education. The reports consistently call for increasing the number, critical thinking skills, and diversity of the students in higher education science and engineering fields.

Informal Science Education Institutions

Informal education settings are another critical piece of the education system. These environments are often thought of as locations for learning that happens outside school across the life span. These environments include museums, science centers, aquariums, zoos, nature centers, libraries, after-school programs, adult community programs, Internet-based learning communities, and TV and radio programming. Informal education participants are diverse and can include learners of all ages, cultural and socioeconomic backgrounds, and abilities. Ideally, educational experiences in informal settings enable learners to connect with their own interests, provide an interactive space for learning, and allow in-depth exploration of current or relevant topics "on demand" (Falk and Dierking, 2000; Griffin, 1998). These types of informal science learning experiences can lead to further inquiry, enjoyment, and a sense that science learning can be personally relevant and rewarding (National Research Council, 2009).

The federal and state governments play an important role in informal education, managing or providing funding for many nature centers, museums, aquariums, zoos, and after-school and adult science learning programs. In addition, the government can shape how these environments interact with the other parts of the education system. For example, there are calls from federal agencies for informal education institutions and other supporters of informal education (e.g., federal government agencies, national foundations, nonprofit research organizations, and advocacy groups) to be active in the reform of the nation's STEM education efforts (National Science Board, 2007).

Many organizations focus on informal education in areas related to NOAA's mission. These groups fund education activities, advocate, provide leadership, and establish organizations for informal education professionals. For example, in the Chesapeake Bay area a number of local organizations collaborate with NOAA on informal education initiatives and advocate for science and environmental education issues, fund education initiatives, and develop education programs.

In summary, many actors are working to improve education in areas related to NOAA's mission across the three parts of the education system highlighted above. In each part of the education system, federal agencies have various responsibilities and are engaged in many education initiatives. The following section details the role of various agencies in the sciences related to NOAA's mission and the factors that determine their roles.

FEDERAL INVOLVEMENT IN OCEANIC, ATMOSPHERIC, CLIMATE, AND ENVIRONMENTAL EDUCATION

Critical Issues

Two issues complicate the federal landscape in which NOAA defines its role in education. First, as mentioned earlier, the fields that underpin NOAA's operational and stewardship mission include overlapping scientific and social science issues. Second, several federal agencies have missions that overlap, and thus the science supported by each agency has themes in common with other agencies.

In the federal government, NOAA's responsibilities include a broad range of multidisciplinary fields (e.g., oceanic, coastal, Great Lakes, climate, atmospheric, and environmental science). In addition, the combination of science and stewardship responsibilities means that its education initiatives need to include the goals of improving environmental literacy and influencing stewardship behaviors. Thus, the agency engages in science education, environmental education, and environmental science education, which have both unique and common features (see Figure 2.1).

"Science is both a body of knowledge that represents current understanding of natural systems and the process whereby that body of knowledge has been established and is being continually extended, refined, and revised" (National Research Council, 2007, p. 26). Both elements are essential to science and science education; however, science learning involves much more than learning content knowledge and process skills. In fact, science learning has been said to include six intertwined strands: developing interest in science, understanding science knowledge, engaging in scientific reasoning, reflecting on science, engaging in science, and identifying with the scientific enterprise (National Research Council, 2009).

Environmental education is directed to help people understand human impacts, to change social behavior and individual behavior, and to affect the decision-making choices of individuals. In effect, environmental education addresses some of the stewardship elements of NOAA's educational mission. Although there is significant overlap between environmental education and science education, the former has goals and practices that extend well beyond the latter, and it would be a mistake to envision environmental education as a subcategory of science education, or vice versa.

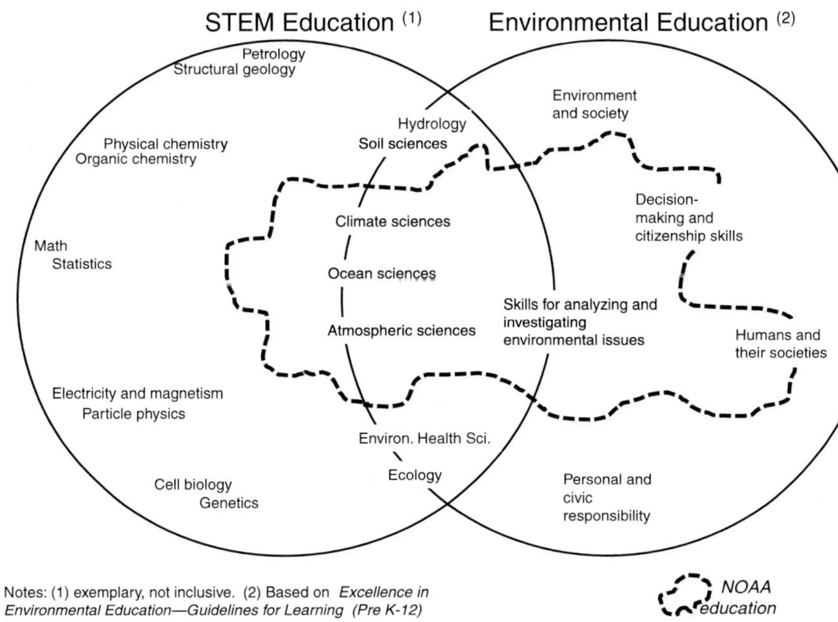

FIGURE 2.1 Relationship among science, environmental education, and NOAA education.

For example, environmental education is rooted in the study of the outdoors, nature, and its conservation. It often focuses on how to change human behavior toward the environment to help preserve nature (e.g., recycling, energy conservation) and the impact of natural systems on humans (e.g., natural resources, recreation). Environmental education by NOAA is a specific type of environmental education, environmental science education. NOAA's environmental education initiatives align with two common approaches to environmental science education. One approach is to infuse the scientific principles of ecology and the interrelationships among organisms and the physical/chemical world (environment) with human impact and decision making. The second approach is concerned less with human impacts and decision making and more with science and the interdisciplinary nature of the study of the environment, spanning the domains of biology, physics, geography, geosciences, chemistry, and mathematics, and uses technology to make various measurements of environmental processes.

A productive way to think about the relationship among science, science education, and environmental education may be to think systemically of a system in which science, science education, and environmental education, working in concert, form a closed loop or feedback system. However, stewardship aspects of environmental science education may sometimes be in conflict with the commerce responsibilities of NOAA (e.g., management of fisheries) and could create potential conflicts across NOAA line offices. Stewardship is also unlikely to align with K-12 STEM learning goals, since these rarely include behavior change.

Coordination of the efforts of the federal agencies that support oceanic, atmospheric, climate, and environmental education has been the subject of many reports, white papers, and discussion documents. These documents have emphasized the need for collaboration, cooperation, and cohesion across the educational efforts of the many federal agencies that support STEM education. This will lead to the avoidance of inefficiency and redundancy and the delivery of a coherent message. Such changes require leadership from within the federal agencies to implement effective changes in the way science, technology, environmental, international, economic, social, and political concepts are conveyed in educational settings.

Each agency has a specific mission, and aspects of all the agency missions overlap and complement each other. Thus, it is not surprising that the education activities the agencies run in support of their missions also overlap and complement each other. Based on their overlapping missions and education efforts, the Academic Competiveness Council (U.S. Department of Education, 2007) has called for coordination and collaboration across agencies that fund STEM education activities; however, such coordination and collaboration can be difficult because of the restrictions on joint funding of programs and incompatible mandates.

Federal Agency Programs

This section is a brief description of the federal agencies, other than NOAA, that have significant oceanic, atmospheric, climate, and environmental education activities, including NSF, the U.S. Department of Education (DoED), the Environmental Protection Agency (EPA), NASA, the National Park Service (NPS), and the U.S. Department of Energy (DOE). We describe DoED and NSF first because they are the primary federal funders of K-12 STEM education programs and research, and their missions are specifically focused on education, although neither agency actually does science or engineering work. We then provide a description of the education activities of the other agencies whose missions include aspects that overlap with NOAA's mission.

This is not a comprehensive list of agencies involved in education in these areas, nor is the description of each agency's investment comprehensive. Instead, these descriptions are meant to be a sketch of the federal portfolio, illustrating that many agencies are involved and that their approaches to supporting education are varied. We relied primarily on the information that the agencies make available on their websites, the information in other National Research Council documents, the expertise of the staff and committee, and the investigative information the committee assembled.

The summary below illustrates that the amount of oceanic, atmospheric, climate, and environmental education supported is agency dependent. Overall the federal agencies value diversity. They have different approaches to build workforce diversity and to reach diverse learners. The approach that each agency takes is unique to the kinds of science it supports. Collectively, the agencies have an opportunity to further support diversity.

National Science Foundation

NSF supports all of the fundamental fields of science and engineering. Unlike other federal science agencies, NSF does not hire its own researchers or scientists or directly operate its own laboratories. Its goal is to identify and support leading researchers and projects to carry out work in areas it deems important. NSF funds a range of education programs, including ones that support education research and development, as well as broader impacts in science, engineering, and mathematics research. The programs reach across the STEM disciplines, and some specifically target oceanic, atmospheric, climate, and environmental education.

The Education and Human Resources Directorate is one of eight NSF directorates, which provides limited-term grants for education research, innovative curriculum development and pedagogy, teacher professional development, education programs and activities, and other educational initiatives. Its budget was about $797 million in 2006, of which about $242 million supported K-12 education research. A small percentage of the grants to principal investigators involve oceanic, atmospheric, climate, and environmental education content.

The NSF science directorates (including geosciences, biological sciences, engineering, and mathematics and physical sciences) support K-12 STEM education through research grants that require recipients to allocate a proportion of the budget to support the "broader impact" related to the research they sponsor. Education activities are one of several activities that provide a broader impact. The Directorate for Geosciences (GEO) is an example of a directorate that supports various relevant education initiatives. For example, it runs a program to broaden participation in the geosciences. In addition, the Division of Ocean Sciences, within GEO, supports Centers for Ocean Science Education Excellence, which is a network of coordinated centers that seeks to support ocean education.

U.S. Department of Education

DoED is one of two federal agencies whose primary mission includes the support of education (NSF is the other). Of its FY2009 budget of $68.6 billion, about $324 million was for education research and development (American Association for the Advancement of Science, 2009). The department's role in education is to establish policies on federal financial aid for education and distribute and monitor these funds, to collect data on U.S. schools and disseminate research, to focus national attention on key education issues, and to prohibit discrimination and ensure equal access to education. Currently, the department's primary influence over K-12 STEM education is through the No Child Left Behind Act and the Education Science Reform Act of 2002. It also provides grants and loans for students to attend college. The department's Institute of Education Sciences has supported education research, but not related to NOAA science or to oceanic, atmospheric, climate, or environmental education content.

DoED supports education in fields critical to NOAA's mission through a new initiative that delivers technical assistance to help five states develop "green" or environmentally friendly career-technical training programs of study. In addition, statewide Math Science Partnerships funded by the department fund districts and schools that train teachers using curricula related to these topics.

Environmental Protection Agency

The mission of EPA is to protect human health and the environment, which gives the agency broad potential ability to contribute to environmental education. EPA was mandated to prioritize environmental education in the National Environmental Education Act (P.L. 101-619), and even though the mandate has expired, it is still the reference document for the agencies' environmental education efforts. The overarching goals of the mandate were for EPA to arrange environmental education initiatives at the federal level and to provide national leadership for the public and private sectors.

EPA provides online education resources on a variety of topics, such as air, waste and recycling, water, conservation, ecosystems, and health and safety. Resources are available for K-12 students and teachers. There is an environmental education working group that works to make web-based environmental education materials accessible, technologically sound, and educationally appropriate.

EPA also provides a number of other programs for educators and students, including distance education, grant programs that support environmental education projects, student fellowship programs, a student award program, informal science education, and other education programs. A few EPA education projects focus on engaging learners of diverse backgrounds. For example, EPA recently formed a partnership with the University of Texas at El Paso to promote a more culturally diverse workforce.

National Aeronautics and Space Administration

NASA's mission is to pioneer in space exploration, scientific discovery, and aeronautics research. It has a multitude of education activities that reach many audiences in formal and informal settings. Its education portfolio includes elementary and secondary education programs, electronic education programs, higher education programs, a national space grant college and fellowship program, and a cluster of university and college programs that support minority research and education. Some programs cover climate, atmospheric, oceanic, and Earth observing issues similar to those addressed by NOAA programs.

An example of an education program that includes aspects of ocean, atmospheric, and climate, and environmental education is the Global Learning and Observation to Benefit the Environment (GLOBE). This is an international program with students from many countries and classrooms collaborating with research investigations on climate change, water conservation, energy use, human health, and other investigations. In addition, NASA recently received funding to support the development of a program called Global Climate Change Education: Research Experiences, Teaching and Learning.

U.S. Department of Energy

The DOE mission to advance the national, economic, and energy security of the United States includes the support of scientific and technological innovation, and the agency ensures the environmental cleanup of the national nuclear weapons complex. One of the agency's overarching goals is to prepare a diverse workforce, including a cadre of diverse middle school and high school master science educators, and a diverse population of students to become scientists, engineers, and mathematicians.

DOE education programs support science learning about clean energy research and technologies. These programs support physical science education at precollege, college, and graduate levels. The DOE also runs the National Science Bowl, a nationwide academic competition that tests middle and high school students' science knowledge in a style similar to the TV show Jeopardy. It also supports the Workforce Development for Teachers and Scientists Program, providing opportunities for K-20 students and teachers to engage in STEM activities: competitions, fellowships, and training.

National Park Service

NPS cares for nearly 400 natural, cultural, and recreational sites across the nation. It helps communities near the parks preserve and enhance important local heritage and close-to-home recreational opportunities. NPS conducts several programs with educational goals and has a separate office for higher education and park initiatives.

The agency engages in a range of place-based education experiences that provide visitors the opportunity to learn about the resources of the national parks. Individual parks support a range of environmental education and stewardship opportunities. Most of the individual parks work with surrounding communities and incorporate regional and local themes of diversity. In addition, the central NPS website provides resources (e.g., lesson plans and datasets) for students and teachers to benefit from field trips to the national parks. A few formal education activities that use the national parks have been supported with NPS as a partner, for example teacher-to-teacher workshops with the Department of Education. Overall, the agency supports a small number of education and stewardship programs, most of which are place-based experiences.

Working Groups, Partnership Programs, and Interagency Collaborations

Existing federal working groups, partnership programs, and interagency collaborations also address education topics related to NOAA's mission. A few examples of these efforts include

- The Committee on Ocean Policy, under the Council on Environmental Quality, which has an Interagency Working Group on Ocean Education.

- The National Ocean Partnership Program, also part of the Committee on Ocean Policy, which seeks to bring coordination to ocean education programs within its 10 federal member agencies (including NOAA).
- The Climate Change Science Program, which integrates federal research on climate and global change, and has supported the *Essential Climate Literacy Principles.*

Federal agencies, education institutions, and private-sector organizations have also collaborated to create literacy principles for ocean science, earth science, and atmospheric science.

THE ROLES OF FEDERAL AGENCIES IN SCIENCE AND ENVIRONMENTAL EDUCATION

With so many agencies contributing to education on topics related to NOAA's mission, it is critical that each agency has a clear role in education and clear responsibilities for education. The role of a federal agency's education program is "grounded in the legislation that defines its individual mission and in the fact that each is an employer of scientists, mathematicians, and engineers and a supporter of the research based in universities and research organizations" (National Research Council, 2008). The role of each agency must be carefully considered by each individual agency in its own way. However, certain aspects of how the different agencies strive to make valid contributions to science and environmental education are common across agencies. This section briefly discusses three of those aspects: the reasons for agency involvement, their resources, and the limitations of their involvement.

Reasons for Agency Involvement

Many federal science agencies are involved in training the next generation of engineers, technologists, and scientists at the graduate level, where supporting education and supporting research are closely connected. Yet the seed for student interest in science, mathematics, technology, and engineering careers can be planted during childhood and cultivated through effective educational opportunities at every stage along the path to a STEM career (Tai et al., 2006).

The involvement of federal agencies in science and environmental education is partially drawn from a commitment to the American public for their support of agency-related science and engineering work. For example, many of the scientific contributions of NOAA, especially in regard to its science missions, are advancements in knowledge—about weather, the ocean, coastal and deep-water resources, and climate, for example. NOAA is committed to making those advancements accessible to the public as well as to other scientists, which requires communicating with and educating the public.

Federal agencies also see a responsibility to support or develop programs that seek to increase the nation's scientific literacy. The goal is not to give each citizen the ability to make judgments about purely scientific issues, but rather to give all citizens enough basic knowledge to be informed participants in public discourse on issues related to science. Although those issues usually also involve questions of economics, ethics, and moral philosophy, among other subjects, knowledge of the basic science is critical. For example, when people have some knowledge of the complex set of scientific factors that influence the Earth's climate, whatever their personal values, it enhances their ability to thoughtfully participate in civic discussions on issues related to alternative fuels and global warming.

In the areas of education related to NOAA's mission, the role of each agency is unique to the federal mandates that shape its resources and audiences. From the perspective of sharing knowledge about science and encouraging stewardship, federal science agencies have three key advantages to support education systems: access to working scientists and engineers, the knowledge generated through their funded programs, and oversight of national resources (National Research Council, 2008).

First, agency-supported scientists and engineers can play an important role in ensuring that education curricula present science, the scientific process, engineering, and the process of design and development in ways that are accurate, up to date, and engaging. Agency education programs can involve their scientists and engineers in modeling the nature of science and engineering and in improving teacher understanding of the science content they teach. Scientists and engineers can also be used as role models for students, sharing their enthusiasm for their work and its challenges and allowing them a real-world glimpse of the possibilities that such careers offer.

The second resource is the agencies' knowledge and support of science and engineering. They support cutting-edge research, engineering, and technology through their grants and contracts as well as through the efforts of their scientific and engineering research staff. Through these efforts they contribute compelling data and ideas that are valuable resources to students, educators, and the public. As public agencies, they have a responsibility to promulgate this information and to make sure that all members of the public, and educators in particular, have access to what they have learned. In some cases, the data themselves can be made available in ways that allow people to interact with them in meaningful scientific investigations and engineering design activities, thereby providing a window on the world of science and engineering that goes far beyond that of classroom investigations and school laboratories.

The third resource is access to the national resources that agencies monitor and protect (National Research Council, 2008). For example, NOAA, EPA, and NPS have responsibilities to monitor and protect some of the country's most delicate and important natural resources. With the exception of some very remote locations, the areas under the purview of these agencies are unique environments in which people of all ages can engage in hands-on authentic educational experiences that can support STEM learning and can change how people interact with the natural world.

Limitations of Agency Involvement

Federal agencies have limited means to support education and research on teaching and learning. In all their educational efforts, agencies need to be informed by the best available knowledge about what is effective in education and how their programs contribute to a larger national education effort. The best way for agencies, including NOAA, to achieve this marriage of science and education is to use and encourage connections between the science, engineering, and education expertise inside and outside the agency.

Universities and academic research organizations as well as organizations of professional educators, such as the National Science Teachers Association and the Association of Science and Technology Centers, are key resources for knowledge of the specific needs of classroom teachers and planetariums, nature centers, aquariums, museums, and science centers. Experienced curriculum development and professional development organizations know how to produce and disseminate educational materials that are both effective and compatible with national and state science education standards. State and local STEM education leaders can provide knowledge about the regional education systems, standards, and needs. Agencies need to be attentive to opportunities to develop contractual or partnership relationships that build on the expertise of people and groups knowledgeable about science and environmental education.

The role of federal science agencies in education is also limited in terms of the breadth and depth of the initiatives that they can undertake. These agencies are not in a position to independently develop and support programs that affect teaching practices, student learning, or systemic reform at a level that would result in national change. Even efforts by the largest federal agencies are dwarfed by the number and variety of school systems in the United States. Agency projects are therefore faced with striking a difficult balance between trying to make a broad impact and providing meaningful engagement on a smaller scale. This balance can be mediated through modern technology, such as the Internet, which can be used as a distribution tool, and through strategic partnering with other science and education agencies and organizations. Agency activities to address the needs of the education system should give priority to the needs of a nation and the local communities in which it works. Therefore, as an overarching rule, federal agencies should strive to understand and complement what schools, school districts, states, informal education institutions, and other actors in the national education landscape are doing. The assets of the agencies can complement ongoing initiatives in the education landscapes and support initiatives that could not take place without their help.

The role of federal agencies is also limited by the fact that programs must be matched to the primary mission of the agency. If an agency embarks on education programs that have little or nothing to do with its mission, it will be acting without expertise or authority. The resulting project is unlikely to be sustained.

Agencies' education programs are also limited by the lack of control they have over the direction of the larger mission and funding. Shifts in the primary mission of an agency as a whole can affect its education programs, especially those with long-term objectives. Radical shifts from changes in administration or changes in agency priorities can result in lack of stability in education programs and erratic funding, which can seriously diminish the program's effectiveness. A funding pattern that fluctuates with federal and agency priorities can hamper the development and maintenance of effective education work—in NOAA and in other federal science agencies.

The reasons and resources for agency engagement and limitations to that engagement apply to the federal agencies that support K-12 STEM education. How the agencies address their role, use their resources, and address their limitations varies depending on their mission, expertise, and funding. Next we take a closer look at where NOAA fits into the landscape of federal involvement in education.

Factors That Define NOAA's Role

NOAA's mandate includes research, education, and stewardship. This provides a unique opportunity for information to flow among these activities—for example, new scientific discoveries can flow into education activities, and questions about effective stewardship can influence research priorities. NOAA's role is further defined by the areas of need in science and environmental education that align to its mission and mandates, such as broadening participation and interest in the related fields. This section describes the unique attributes that the agency can bring to education as well as the areas of need that the agency seems well positioned to address.

Assets for Supporting Education

Place-based learning environments have been shown to be well suited for increasing interest in and understanding of the environment for individuals of all ages (National Research Council, 2009). The educational opportunities afforded by the natural resources managed and protected by NOAA (often in collaboration with other federal and state agencies) are vast and important. Management of such places provides the agency with connections to the surrounding communities and organizations concerned with environmental issues—connections that can be used to create partnerships with states, localities, and other organizations. Such partnerships and coordination with other federal, state, and local entities that manage similar areas for education programs can engage learners with hands-on participation in science and give them opportunities to witness science taking place, as a core part of the activities.

Another major asset that NOAA brings to education is the science, engineering, and technology that it supports and produces. The agency has access to scientists and engineers doing cutting-edge work in oceanic, atmospheric, climate, and environmental science, both within NOAA and through its support of external scientists and engineers by way of grants, cooperative agreements, and contracts. These endeavors involve a range of technology and scientific resources to investigate the Earth's atmosphere, oceans, and climates, including research ships and remotely operated vehicles, satellite systems, data systems, and technology that is sometimes jointly used by other scientific agencies. Technologies, scientific resources, and data systems can provide opportunities for students and citizens to see scientists in action and participate in research, data collection, and analysis.

Areas of Need in Which NOAA Can Contribute

There are five areas of need to which NOAA can contribute, although it does not have to act alone. NOAA needs to work in collaboration with appropriate partners to effectively contribute to the nation's education systems.

1. *Integration of science, education, and stewardship.* The science, education, and stewardship goals that guide NOAA's mission are interconnected. In the United States, very few organizations can lay claim to expertise and authority in all three of these domains. NOAA has the potential to support an integrated feedback loop by leveraging its collaborations with the science, engineering, education, and stewardship communities. For example, NOAA personnel and programs could partner

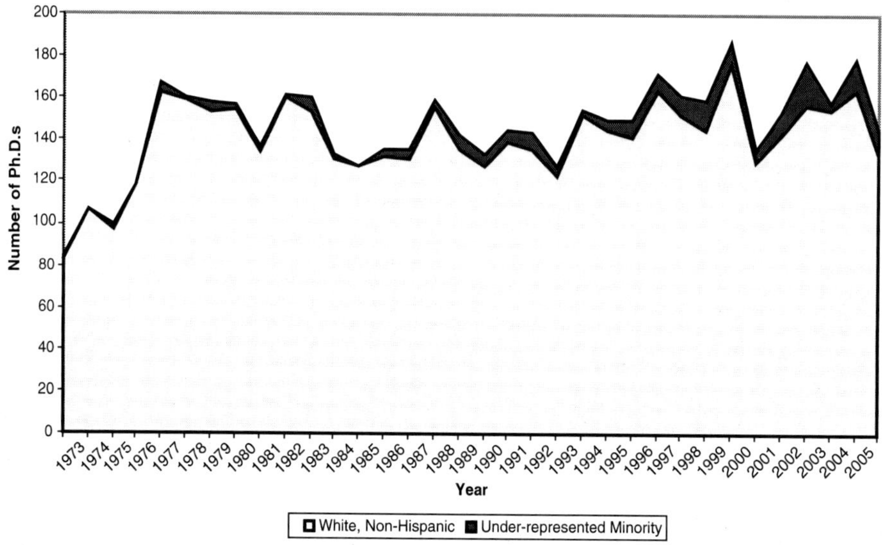

FIGURE 2.2 Ph.D.s in atmospheric and oceanic sciences earned by U.S. citizens and permanent residents.
SOURCE: Adapted from National Science Foundation (2010).

with the academic research community, industry, and states (schools, state agencies) to gather data about the oceans and atmosphere through systematic observations, convert these data into understandings about process via modeling and other interpretive techniques, and develop education materials and train teachers to convey these scientific understandings to learners of all ages. In a connected and parallel effort, NOAA could work with partners to develop stewardship programs to encourage environmentally sustainable behaviors and decisions, as well as interest in and input into scientific research priorities.

2. *Diversity in the ocean, atmospheric, and climate science workforce.* The rate of minority participation in the atmospheric and oceanic sciences is lower than in many other sciences, in spite of several well-intentioned and ongoing programs (see Figure 2.2). Multiple reports highlight the continuing lack of diversity among degree seekers, recipients, and the workforce in fields related to NOAA's mission (Levine, González, and Martínez-Sussmann, 2009). A recent National Research Council report (2009) points out:

> The challenges in engaging nondominant groups in the sciences are reflected in studies showing that (1) inadequate science instruction exists in most elementary schools, especially those serving children from low-income and rural areas; (2) girls often do not identify strongly with science or science careers; (3) students from nondominant groups perform lower on standardized measures of science achievement than their peers; (4) although the number of individuals with disabilities pursuing postsecondary education has increased, few pursue academic careers in science or engineering; and (5) learning science can be especially challenging for all learners because of the specialized language involved.

Several lines of reasoning are the basis for supporting participation of individuals from underrepresented populations. The demographics of the United States are changing. By 2039, the working-age population will include over 50 percent of current ethnic minorities (U.S. Census Bureau, 2008). Success in workforce development therefore depends on attracting and retaining individuals from minority backgrounds. Equally important in a service-oriented agency like NOAA is that its ability to serve the population depends on having a workforce that reflects it. Others argue that diverse groups should be served because in a democratic society all people are equally entitled to a quality education and to the knowledge that will enable them to participate in civic decisions, including an increasing number of decisions related to the environment. In fact, the idea of environmental justice suggests that minority participation in and knowledge of environmental sciences is a tool for ensuring that minority communities do not suffer disproportionately in decisions about the use of environmental resources and the disposal of pollutants. Finally, there is also an argument that having a diverse scientific workforce has a positive benefit on the outcomes accomplished by science. When diverse perspectives are included in science, different views of the natural world are introduced, different interpretations of findings are made, different questions are asked, and different methods are used (Bang, Medin, and Atran, 2007; Fox Keller, 1995; Longino, 2003; Louis, 2007). For example, when women began to enter the field of women's health, different scientific questions were posed, and advancements in the field followed (Longino, 2003).

The especially low rate of minority participation in the field of geosciences (which includes the oceanic, atmospheric, climate, and environmental education categories) raises a number of questions regarding whether unique factors are involved in that field. Levine, González, and Martínez-Sussman (2009) reviewed the published literature and identified factors that may be impacting the geosciences:

- The small size of the field means many people do not know the field or a role model in the field (Huntoon and Lane, 2007).
- In many school systems, geosciences are not taught at the high school level.
- Students in schools with high poverty levels or high minority percentages are less likely to take mathematical prerequisites for these sciences, such as trigonometry, advanced algebra, and pre-calculus (National Science Board, 2009) and are more likely to be taught mathematics by a teacher whose specialization is not mathematics (Education Trust, 2008).
- Cultural and religious beliefs may conflict with geoscience practices or ways of interacting (National Science Foundation, 1996; Seymour and Hewitt, 1997). For example, reductionism in science and association with mining is a negative factor for some Native Americans.
- Obligations to serve the community in more specific ways may be higher in certain underrepresented communities (Seymour and Hewitt, 1997).
- Competitive emphasis in geosciences (and other sciences) may be more at odds with cultural practices in minority communities (Seymour and Hewitt, 1997).

3. *Quality and quantity of oceanic, atmospheric, and climate education across the country.* In general, ocean sciences are only sparsely represented in K-12 science education standards. This is true of the national standards and of the state standards, even in many coastal states. In atmospheric sciences, weather is the one topic that is widely taught, but climate, including climate change (Kastens and Turrin, 2008), has not yet found its way into the standards of many states.

As mentioned above, NOAA's education programs must align with the existing education standards in the localities in which they operate. However, NOAA can also play a useful role in major science education efforts, such as the new National Assessment of Educational Progress in science, the revision of the *National Science Education Standards* (also known as common core science standards), as well as revisions to state education standards in states in which it has facilities. Scientists sponsored by NOAA could serve as content experts, and education staff supported by NOAA could provide exemplary materials and assist in other ways. As a federal agency, NOAA would have to be careful to respect the state's primacy in K-12 education; in many ways this can be done by using local academic research and education resources. Given how full the K-12 science curriculum is already, standards revision efforts should look for opportunities to intertwine oceanic, atmospheric, and climate concepts throughout the core sciences, for example by contrasting marine and terrestrial food webs in biology, or considering the mechanism of the greenhouse effect in the physical sciences. Concepts from the oceanic, atmospheric, and climate sciences may also provide exciting, nearly tangible, and locally relevant examples of "big idea" or core concepts common to many disciplines (core concepts in science are discussed further in Chapter 3). Efforts to inform standards should be coordinated with other federal and state agencies in promoting these changes.

4. *Field-based education.* Field-based learning experiences offer personal experiences and direct contact with the actual environment and have been shown to contribute to people's understanding of and commitment to environmental conservation and stewardship (Bogner, 1998; Dillon et al., 2006). The need for field-based education in formal educational settings is becoming stronger with the passing years, as many urban and suburban children spend little of their free time in unstructured outdoor play or exploration in natural settings (National Research Council, 2009). Field-based education is a critical component of oceanic, atmospheric, and environmental education. Yet it is one aspect of education that schools find hard to do well.

To optimize field-based learning experiences, the education staff at each NOAA-sponsored venue must work closely with the teachers and administrators of regional schools to build a relationship of trust and ensure that the field experience is well integrated with the curriculum. Learning goals should be clearly articulated, with a purposeful balance between science and stewardship goals. In addition to striving to accommodate the schools that come to them, NOAA-supported field sites should proactively reach out to underresourced schools and schools with populations that are underrepresented in science and engineering.

5. *Coordination across federal agencies.* With so many players involved in science and environmental education at the federal level, there has been a proliferation of differing goals and standards, different methodologies, and even different values, norms, and cultures. Education initiatives are rarely coordinated across agencies, and cross-agency cooperation is limited. Collaboration, coordination, efficiency, and cohesion are needed to limit redundancy across the science and environmental education that NOAA is involved in and to ensure that teachers and learners get a productive and complementary view of the science. To take on such a critical role, NOAA is likely to need both fiscal and legislative support. This need to coordinate federal agency activities and address interagency disputes is highlighted in reports from the Pew Ocean Commission (2003) and the U.S. Commission on Ocean Policy (2004).

Coordination of education efforts entails a high level of coordination both within NOAA and with other agencies, particularly with regard to the sharing of promising practices, awareness of education initiatives at each agency, and promoting partnerships. The range of fields that NOAA supports through its scientific and engineering endeavors makes the agency well suited to develop education initiatives and also to support coordination and cohesion across agency education projects. NOAA has taken on coordinating roles on interagency education groups, including the Interagency Working Group on Ocean Education and the Climate Change Science Program's informal interagency climate outreach and communications group. NOAA could take a leadership role in other areas of earth systems education, in which such coordination does not yet exist.

Other agencies may also have the capability of taking a leadership role in these areas, but we did not carefully review the assets of other federal science agencies. We think NOAA has the needed assets and capabilities to be one of the federal agencies that take a leadership role in coordinating federal education efforts. In climate, for example, NOAA (possibly along with other agencies) could step up to support the call for interagency coordination laid out in the U.S. Climate Change Science Program. If NOAA can convincingly integrate its two mandated sciences, atmospheric science and ocean science, then this could be a model for how to reach out to the many agencies that support learning in all of the earth sciences. As it does so, it may be able to draw on the more interdisciplinary approach of its regionally focused education efforts, like the marine sanctuaries, Sea Grant colleges, and the estuarine research reserves. In addition, many natural settings are protected, managed, and maintained by other federal, state, and local agencies and tribes that are similar to those protected, managed, and maintained by NOAA. Coordination between NOAA and the other federal, state, and local agencies and tribes would allow materials and priorities to be streamlined.

3

The Education Portfolio and Effective Practices

The National Oceanic and Atmospheric Administration (NOAA) released its education strategic plan for 2009-2029 during our review of the agency's education programs. The related implementation plan was available only after our study concluded, so it is impossible for the committee to judge the impact or the implementation of the education strategic plan. We are able to critique the strengths and weaknesses of the plan and provide guidance on how to improve it. In general, the plan provides a very high-level description of how education programs are managed and their goals, topics to be addressed in reaching their goals, the outcomes related to the goals, and strategies for reaching the outcomes. In this chapter we describe how education is managed and funded at NOAA, critique the strategic plan, and summarize effective education practices that can guide the implementation of the strategic plan.

PORTFOLIO MANAGEMENT

NOAA is a decentralized organization, and each of its six line offices, incorporated at different times through separate mandates or internal initiatives, has a different mission (see Box 1.3). As an artifact of the agency's decentralized nature, education programs have developed in a decentralized manner across five of the line offices and the recently created (in 2003) Office of Education.

Five of the line offices manage education efforts:

1. the Office of Oceanic and Atmospheric Research (OAR),
2. the National Ocean Service (NOS),
3. the National Marine Fisheries Service (NMFS),
4. the National Environmental Satellite, Data, and Information Service (NESDIS), and
5. the National Weather Service (NWS).

Education programs are also managed by the Office of Education, a corporate function in NOAA's organizational structure.

The efforts of these offices are coordinated by the Education Council (EC), which was created in 2003 at the recommendation of the Subcommittee on Education of NOAA's Scientific Advisory Board. The EC monitors and evaluates education programs across the agency and provides recommendations regarding their conduct and guidance for their future direction. It serves as an internal forum for the discussion of ideas and proposals for NOAA-wide education and outreach activities and priorities and for making recommendations to NOAA management on all aspects of educational activities.[1] The EC guided the development of the education strategic plan, providing input on educational goals, outcomes, and strategies. It served a similar role in developing the education implementation plan.

All of NOAA's education programs are represented on the EC, with the Office of Education having the greatest amount of responsibility. The EC chair is the director of the Office of Education, the vice chair is the deputy director, and the executive secretariat is the Office of Education. The principal members of the EC are as follows:

- Director, Educational Partnership Program, Office of Education
- National education coordinator, Office of the National Marine Sanctuaries, NOS
- Education program leader, National Sea Grant Program, OAR
- National education coordinator, National Estuarine Research Reserve Program, NOS
- National education coordinator, Ocean Exploration Program, OAR
- National education coordinator, NOAA Corals Protection and Conservation Program
- Climate education coordinator, Climate Program Office, OAR
- Education and Teacher at Sea program manager, NMFS
- OAR (at large)
- NOS (at large)

[1]See http://www.oesd.noaa.gov/council/meetings.html for Education Council meeting agendas [accessed May 2010].

- NMFS (at large)
- NWS (at large)
- NESDIS (at large)

Representatives from the NOAA Central Library, Office of Program Planning and Integration, Office of Program Analysis and Evaluation, Legislative Affairs, Office of Communications, Ecosystem Goal Team, and Weather and Water Goal Team serve as advisers to the EC.

Each principal member of the EC (but not the advisers) has a vote in its decisions. Although the group strives for consensus, the final decisions are made by the chair, who maintains 51 percent of the vote. It is unclear how the EC enforces decisions when consensus is not reached. Without the cooperation of the members from the 15 line and program offices, the EC, as a body, has no clear power to enforce its decisions.

Recently, NOAA created the Executive Council on Engagement (ECE). The ECE is comprised of the chair of the EC, the director of communications, the chair of the Regional Collaboration Executive Oversight Group, and the chair of the Extension and Training Services Committee. The ECE provides guidance and recommends actions to the NOAA Executive Panel (NEP) to promote a dialogue and a two-way relationship with the public to identify, develop, and improve products and services to meet society's needs. The focus of this group is on extension, communication, regional collaborations, and education. The chair of the EC works with the ECE to ensure coordination across these efforts. The ECE was created as our study was being finalized, and thus very little information on its impact or functionality was available.

Individual offices have separate mandates, and some have local components with local control. Education programs are managed differently by the line and program offices and the Office of Education as a result of available resources for education (staff and funding) and mandating language. For example, the Office of Education has a small centralized staff in the Washington, DC, metropolitan area and does not carry out or fund scientific research, exploration, or stewardship activities, as the line offices do. Thus, the majority of the Office of Education's programs are run by partners external to NOAA (although some grant recipients are internal). In contrast, the education coordinators from the line and program offices on the EC work with their education program staff across the country to implement education activities, which are based on the scientific, exploration, and stewardship activities of the line and program offices.

Each line and program office coordinates and oversees its education staff and programs in different ways, some of which can be quite complex. For example, the NWS staff respond to local requests for education activities, yet they must prioritize their research and warning coordination responsibilities ahead of their education responsibilities. This limits the managerial influence of the national NWS education coordinator. The management of education in the National Estuary Research Reserve System (NERRS) is particularly complex. Each NERRS site is a partnership between NOAA and the state government. Both NOAA and the state fund some of the education staff; in some cases, each funds a portion of individual staff members. This blending of support within reserves means that education staff answer not only to NOAA but also to the state. For example, at Elkhorn Slough National Estuary Research Reserve in California, the education coordinator is funded primarily by the California Department of Fish and Game, and the coordinator answers to the reserve manager, who is also funded by the Department of Fish and Game. Yet two of the three education staff under the coordinator are funded by NOAA.

BUDGET

The goals of NOAA education are formidable, yet the budget is small: \$43 million for education in 2008. Within the federal government, NOAA's portion of the funding for science education is also small. For example, the Department of Education, the National Science Foundation, and the Department of Health and Human Services account for over 80 percent of science education spending by the federal government (see Figure 3.1). These departments combined spend approximately \$2.5 billion on science education per year on programs that include graduate fellowships, mathematics and science partnership grants, and undergraduate financial aid. The National Aeronautics and Space Administration, one of the largest education funders among federal science mission agencies, had an education budget of approximately \$162 million in 2006 (U.S. Department of Education, 2007). In contrast, NOAA's education budget in 2006 was less than a quarter of this (U.S. Department of Education, 2007).[2]

NOAA's overall education budget has remained relatively consistent fiscal year (FY) 2005 to FY2008 (see Figure 3.2). NOAA's education budget decisions are often influenced by congressionally mandated appropriations (known as earmarks) and mandates, putting added constraints on how education is managed. Figure 3.3 illustrates how the education budget breaks down across programs, providing a general sense of the size of the education budgets for all of NOAA's education programs. Although only three programs focus on higher education (the Educational Partnership Program; Hollings and Foster Scholarship Programs), they account for about 33 per-

[2]NOAA's education budget in 2006 is reported as \$39 million in the ACC report (U.S. Department of Education, 2007); however, in files received by the committee, it is reported as approximately \$60 million.

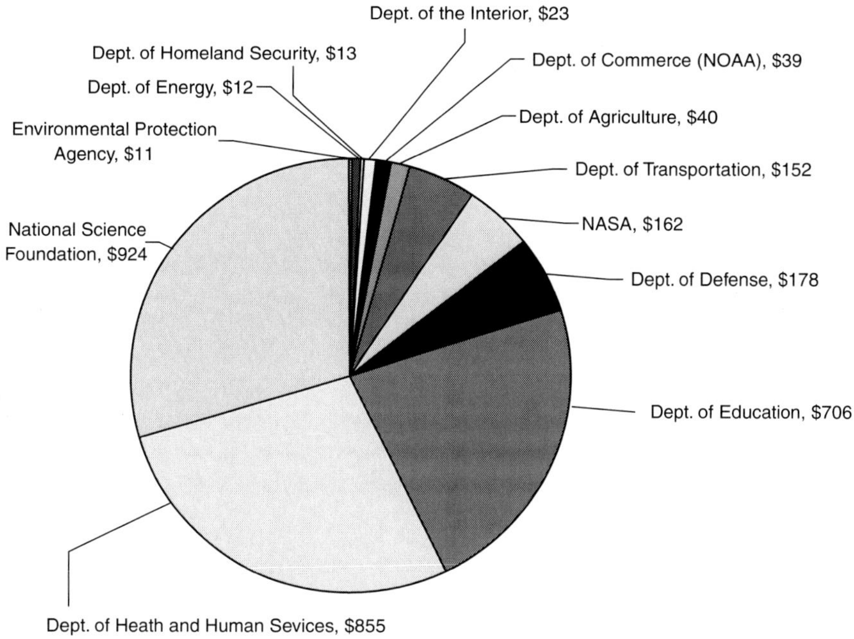

FIGURE 3.1 Federal 2006 education budget ($ in millions).
SOURCE: U.S. Department of Education (2007).

cent of the budget. In addition, much of the Sea Grant budget is for higher education, and the remainder supports various formal and informal K-12 and continuing education activities. Beyond this rough separation between higher education and environmental literacy programs, it is challenging to make any finer budget distinctions.

It is very difficult to get a complete picture of the full measure of NOAA's engagement in education by analyzing the budget, because one cannot simply follow the money to find the education activities. During testimony and during informal site visits, we encountered scientists, extension personnel, and educators who indicated that they are engaged in the education mission but that it does not really show up in their job portfolios, program reviews, or budgets. In addition, some programs support education activities but do not report a separate education budget category (e.g., the cooperative institutes). This means that it is very difficult to discern the full footprint of NOAA's education activities.

Even when it is possible to track the amount of education funding in a program (e.g., the Office of the National Marine Sanctuaries

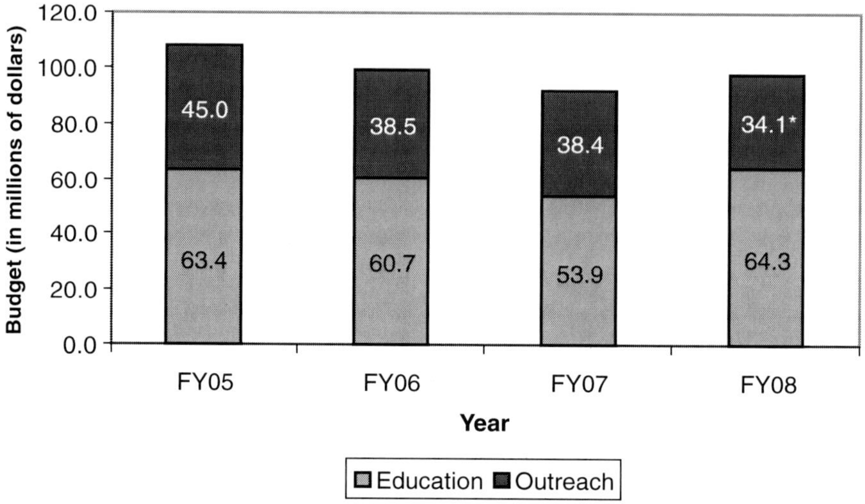

FIGURE 3.2 NOAA education and outreach budget, 2005-2008.
NOTE: *2008 information is based on the projected budget.
SOURCE: NOAA Education Council.

Program, Teacher at Sea, the National Estuarine Research Reserve System), it is difficult to track how much funding is provided to individual activities run at the local level, because we were not able to locate a document that breaks down the budget in this manner. For example, the National Sea Grant College Program's $5.4 million for education and additional $16 million for outreach are allocated across the 32 universities and their partner institutions. Each institution has developed education and outreach activities with its funds, yet the education and outreach budgets of each of the Sea Grant colleges were not readily available. Furthermore, the cost share and matching funds that some NOAA education programs receive from other agencies, organizations, and institutions—including the Office of National Marine Sanctuaries, the National Estuarine Research Reserve System, and the Bay-Watershed Education and Training Program—make it difficult to decipher NOAA's investment in the education programs it supports. Each of these factors makes it difficult to assess how much money is going toward teacher professional development, curricular support materials, informal education activities, or other education activities. What is clear is that NOAA will need to continue to develop and nurture partnerships if it is going to achieve its ambitious strategic plan.

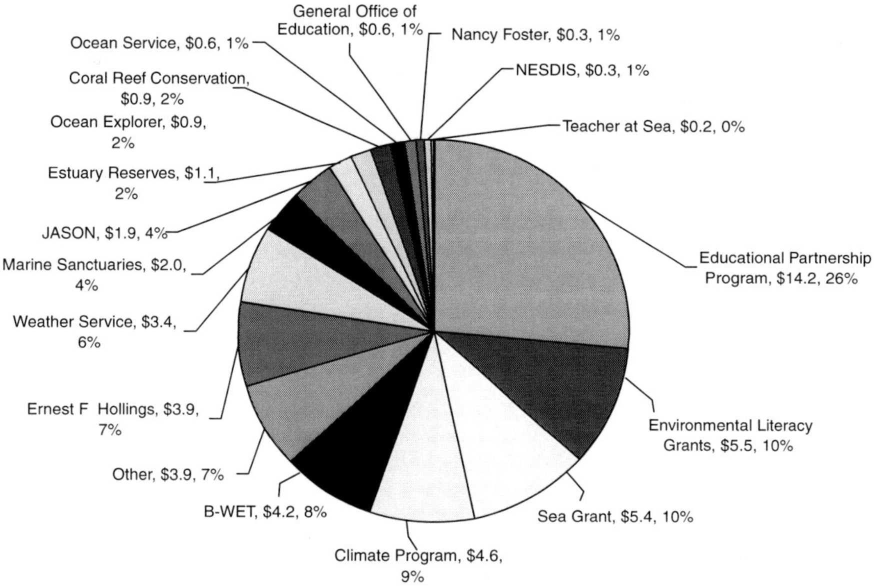

FIGURE 3.3 NOAA 2007 budgets for education ($53.9 million).
NOTE: The "Other" category was created by the committee for monies that we were unable to assign to one of the major programs. Budget items that fall into this category include the Coastal Services Center, the National Centers for Coastal Ocean Science, Habitat, the Satellite Acquisitions and Satellite Service, Geodesy, Aviation Weather, Coastal Estuaries and Oceans, Hydrology, Air Quality, Space Weather, and Science, Technology, and Infusion.
SOURCE: NOAA Education and Outreach spreadsheet.

EDUCATION STRATEGIC PLAN

The 2009-2029 education strategic plan, which updates the 2004 education strategic plan, was developed to create greater coherence and cohesion across the agency's education portfolio.[3]

NOAA's 2009-2029 education strategic plan updates the 2004 education strategic plan. Under the mandate of the America COMPETES Act, NOAA must revise its education plan every five years. The document reflects the collaborative efforts of the Education Council as well as input from the broader education and resource management community and other interested groups.

[3]The education strategic plan is available at http://www.education.noaa.gov/plan/index.html [accessed May 2010].

Vision, Mission, and Goals

"An informed society that uses a comprehensive understanding of the role of the ocean, coasts, and atmosphere in the global ecosystem to make the best social and economic decisions" is the agency's vision. Its education mission is "to advance environmental literacy and promote a diverse workforce in ocean, coastal, Great Lakes, weather, and climate sciences, encouraging stewardship and increasing informed decision making for the Nation." The agency staff set two primary goals to accomplish this mission:

> *Goal 1:* Environmental Literacy: An environmentally literate public supported by a continuum of lifelong formal and informal education and outreach opportunities in ocean, coastal, Great Lakes, weather, and climate sciences. Environmental literacy is defined in the plan as "a fundamental understanding of the systems of the natural world, the relationships and interactions between the living and non-living environment, and the ability to understand and utilize scientific evidence to make informed decisions regarding environmental issues." NOAA believes that an environmentally literate public is critical to achieve its mission goals related to stewardship, resource management, and preparation for and response to severe weather.
>
> *Goal 2*: Workforce Development: A future workforce, reflecting the diversity of the Nation, skilled in science, technology, engineering, mathematics, and other disciplines critical to NOAA's mission. This goal benefits the agency and helps to build a workforce with the skills that are crucial to maintaining America's competitiveness and ability to collaborate internationally.

The plan lists six outcomes related to Goal 1 (listed in Box 3.1), which are described as interdependent actions that NOAA will pursue to accomplish the environmental literacy goal. NOAA believes that the environmentally literate public supported through these actions will also provide a base for a continuous supply of the nation's future workforce, which aligns with research that shows interest in science, technology, engineering, and mathematics (STEM) careers develops much earlier than college (Tai et al., 2006). Four topics are cited as integral to these outcomes: promoting environmental stewardship, facilitating change in education systems, connecting citizens to nature and community, and using emerging technologies.

The outcomes related to Goal 2 (also listed in Box 3.1) are framed in terms of needs within NOAA, yet the plan also states that it is "committed to developing the Nation's workforce beyond the specific needs of NOAA." This goal and supporting outcomes outline a strategy that the agency will employ to extend the current education and recruitment pipeline to meet the

BOX 3.1
Environmental Literacy and Workforce Development Outcomes

Environmental Literacy Outcomes

Evaluating Education—NOAA education programs are developed and refined using the best available research on the effectiveness of environmental and science education

Literacy Principles—Educators understand and use environmental literacy principles

Inquiry Based Learning—Educators, students and/or the public collect and use ocean, coastal, Great Lakes, weather, and climate data in inquiry and evidence-based activities

Lifelong Learners—Lifelong learners are proved with informal science education opportunities focused on ocean, coastal, Great Lakes, weather, and climate topics

Partnerships—NOAA works cooperatively to maximize the impact of federal investment in ocean, coastal, Great Lakes, weather and climate education

Engaging Audiences—NOAA's education community functions in a unified manner and is coordinated with agency extension, training, outreach and communications programs to fully engage NOAA audiences

Workforce Development Outcomes

Career Development—A diverse and qualified pool of applicants, particularly from underrepresented groups, pursues student and professional opportunities for career development in NOAA mission-critical disciplines

Programs and Activities—NOAA's employees support programs and activities for students and teachers to learn about and explore NOAA science and stewardship

Career Paths—A diverse pool of students with degrees in STEM and other fields critical to NOAA's mission connect to career paths at NOAA and in related organizations

SOURCE: National Oceanic and Atmospheric Administration (2009).

national workforce needs of tomorrow. Three topics are highlighted as integral to accomplishing these outcomes: workforce development for students, educators, researchers, and managers; support for NOAA mission-critical disciplines (not just science); and support for underrepresented populations in sciences related to NOAA's mission.

The plan lists three or four strategies for reaching each of the outcomes under both goals. The strategies are short, concrete statements about how education initiatives can be designed and implemented to accomplish the stated outcomes. However, more detailed guidance is expected in the coming implementation plan.

Critique of the Education Strategic Plan

The strategic plan is an important sign of progress and is especially valuable in helping to make education a coherent priority across the agency. The current plan provides more detailed description of the expected outcomes than previous ones, and it outlines strategies to bring about those outcomes.

The overarching goals set forth in the strategic plan align with NOAA's strengths: involvement in marine, climate, and atmospheric scientific and engineering research and conservation/protection of the nation's marine resources. The agency's education programs bring its research findings and conservation work to the public and in the process also encourage individuals to pursue careers in these fields. The program goals are in line with the nation's need for a scientifically literate populace, as well as its workforce needs. The awareness that activities should be based on effective practices in education research, as well as a commitment to contribute to education research, are excellent steps forward for NOAA education.

One positive and critical aspect of the 2009 plan is the brief summary of the agency's desired role in the national landscape of ocean, coastal, Great Lakes, weather, and climate science education. In the partnership and collaboration section, NOAA states that it is the leading science and service agency in ocean and atmospheric science, and thus it has the responsibility to increase its role as a coordinator and collaborator in these areas of science education. The America COMPETES Act is cited as a mandate to serve as a "catalyst" to strengthen oceanic and atmospheric science. The strategic plan points to a broad array of potential partners, including other agencies, businesses, organizations, professional societies, education associations, and school systems. All partnerships will be developed in the interest of the agency using its resources to "advance the environmental literacy and scientific knowledge of our Nation and the global community." Partnerships will be critical if NOAA is to reach the ambitious goals identified in the strategic plan, because the agency does not have the resources to achieve its goals on its own.

The strengths of the plan also include an emphasis on including more members of historically underrepresented groups in fields critical to NOAA's mission, as well as an emphasis on the use of ocean, coastal, and other place-based resources as unique and valuable assets for learning. In addition, the place-based resources of the agency can be used to provide students and the public with hands-on, authentic education activities. The importance of these types of activities is described in the second half of this chapter.

The current education strategic plan is a significant improvement over the 2004 plan, yet it also has weaknesses. First, two issues concern the outcomes defined under the environmental literacy goal and the workforce goal (see Box 3.1). One is that only three of the six environmental literacy outcomes (literacy principles, inquiry-based learning, and lifelong learners), and two of the three workforce development outcomes (career development and career paths) are concrete, measureable constructs that would be expected as a result of efforts to reach either goal. The other outcomes are general statements of intent or strategies to be implemented. An example is the outcome of the literacy goal—the use of research on the effectiveness of environmental and science education to develop and refine education programs. The committee supports NOAA's effort to develop and refine its programs based on effective practices; however, this is not an outcome of providing education programs to promote environmental literacy. Instead, it is more appropriate to frame this practice as an overarching principle that guides the development of educational materials. All outcomes should be concrete measureable constructs that would be expected as a result of efforts to reach either of the two goals.

The second issue is that some critical outcomes are not included. Diversity and broadening participation are central in the workforce development outcomes, but they are only tangentially mentioned as one aspect of the engaging audience's outcome under the environmental literacy goal. An outcome focused on reaching out to underserved populations in the environmental literacy goal should be included to address the national need to expand understanding of and interest in the science and stewardship issues related to NOAA's mission among K-12 students, adults, and the public.

Another weakness: although it states that promoting stewardship is an integral topic to be addressed in reaching its environmental literacy goal, the plan does not include an outcome related to stewardship. Many of NOAA's place-based education activities were developed to promote stewardship behaviors. Thus, there seems to be a mismatch in the focus on stewardship in the plan and the focus on stewardship in NOAA's education activities. Reframing stewardship as an outcome under the environmental literacy framework is one possible mechanism to address this mismatch.

The education strategic plan does not include critical details regarding how it will accomplish the two goals. For example, there is no clear articulation of the workforce development needs in the mission-critical disciplines listed in the plan. It is unclear how NOAA can accomplish its goal of supporting the creation of a "world-class" workforce without a clear understanding of its and the nation's workforce needs in these areas. Although workforce needs are difficult to predict in general, it is particularly difficult to do so in interdisciplinary areas, such as those critical to NOAA's mission.

The agency is nevertheless in need of a clearer estimate of workforce needs to guide the scope and direction of workforce-supporting programming.

The education plan does not discuss how it will use the education assets within and external to the agency. First, there is no mechanism to bring the local education staff from different programs together to share effective practices across education programs. The plan mentions the need for internal coordination to support education activities; however, the coordination being discussed is at the EC level. Promoting connections among local education staff can be just as valuable in creating internal coordination.

Second, although the plan highlights the use of technology to support science education, it does not mention the use of technology to support engineering education. Overall, there is very little mention of engineering education in the education plan, even though it is a critical aspect of the research, exploration, and stewardship work related to NOAA's mission. NOAA and its partners have a wealth of engineering expertise that could be leveraged in the agency's education activities. A recent report of the National Academy of Engineering and the National Research Council on engineering and K-12 education has explored the subject of the current status of incorporating engineering into K-12 education and what should be done to improve the situation (National Academy of Engineering and National Research Council, 2009).

The scientists, engineers, and other experts within and external to NOAA are another asset that the plan does not clearly provide a role for. Many of them are engaged in outreach and education work and have passion and knowledge that can contribute to educational activities. By clearly identifying responsibilities for these professionals, the plan could ensure that they have a role to play that they understand.

Two terms used in the education strategic plan are potentially confusing. The term "NOAA science," used throughout the plan, is vague and should be reconsidered. The confusion with this term is expanded on in Box 3.2. The plan also includes a set of seven principles that should be characteristic of all of the agency's programs. In the plan, NOAA calls these seven principles its "education standards." This terminology can be confusing, given the common use of the term to refer to something entirely different. Also, two of the standards (basing programs on the best science available and continual evaluation and improvement) are redundant with the first outcome under the environmental literacy goal.

EFFECTIVE EDUCATION PRACTICES

There is a growing body of research related to aspects of education that are highlighted under the environmental literacy goal in the strategic education plan, including teaching and learning the sciences related to

BOX 3.2
"NOAA Science" in the Education Strategic Plan

The committee identified at least three possible interpretations of the term "NOAA science," which is used throughout the *NOAA Education Strategic Plan 2009-2029*: (1) the research, or the results of research, conducted by scientists employed by NOAA; (2) the research, or the results of research, funded by NOAA internally or externally; and (3) the body of knowledge and research in the fields related to NOAA's mission, that is, research related to oceans, coastal areas, Great Lakes, the atmosphere, climate, and weather.

In fact, the goals of a particular education activity and the participants served may lead to a focus on a particular interpretation. For example, if a program is designed to serve elementary school-age children and to address the goal of environmental literacy, it may be based on a broad and basic body of scientific knowledge drawn from one or more of the fields of science related to NOAA's mission. However, it may be inappropriate to focus only on the knowledge generated by research funded by NOAA. In contrast, a postdoctoral program designed to attract scientists to positions at NOAA may be appropriately focused on cutting-edge research carried out by intramural researchers employed by the agency who act as postdoctoral fellow advisers. Also, one of the roles of NOAA is to inform the public about the research it funds and particularly about exciting breakthroughs. This should be accomplished both by public relations activities (not the subject of this review) and by selected education activities.

The problem with the term "NOAA science" is that it does not distinguish among these interpretations and allows for misinterpretation. The committee is particularly concerned about narrow interpretations that could lead to an exclusive focus on the research funded by NOAA, thereby constraining education activities that should draw on a broader body of scientific knowledge. In addition, we are concerned that the term may encourage a blurred line between activities focused on education and activities that are more appropriately defined as communication or public relations for the agency. This is not to say that the committee observed such practices in existing education programs, but rather that use of the term could justify a move in this direction. Saying "NOAA science" seems to imply a priority for promoting the results of NOAA-funded research rather than on achieving the goals set out in the education strategic plan.

NOAA's mission, developing stewardship behaviors, structuring informal learning environments, supporting teacher professional development, and reaching underserved populations. There is also a body of literature related to aspects of workforce needs that are highlighted under the workforce goal in the strategic education plan, including developing accurate workforce needs assessments and factors in supporting workforce development that can inform future workforce development programs. Lessons learned, summarized below, from the existing research on these topics should be used to inform the implementation of education programs to achieve both the environmental literacy goal and the workforce goal.

Environmental Literacy Programs

Learning and Teaching the Sciences Related to NOAA's Mission

Several inherent attributes of oceanic, atmospheric, and climate sciences contribute to making these disciplines challenging to teach and learn in formal and informal settings. These attributes include but are not limited to the extreme (both large and small) spatial scale of important processes, the interrelated reliance on models and representations rather than actual target phenomena in hands-on activities, the centrality of systems thinking and emergent phenomena, and the importance of nonexperimental modes of inquiry. None of these difficulties is unique to these sciences and none is insurmountable, but NOAA's education leadership needs to be mindful of these inherent challenges and plan for them in the design of programs, instructional materials, and evaluations.

We provide a thumbnail sketch of the findings from the robust literature base about the attributes of the relevant sciences that make them difficult to learn and then provide evidence-based practices for overcoming these difficulties (see Tran, 2009, for a detailed review).

The size of many important ocean and atmospheric phenomena, especially those too large for students to have experienced directly, may make it harder for them to form an accurate mental model. Both students' and teachers' grasp of structures, processes, and phenomena of varying scales is strongest for phenomena closest in size to the human body, and it weakens as the scale becomes large or small relative to the body (Jones et al., 2007b; Tretter et al., 2006). Thus, hands-on instructional activities on environmental topics often are built around models of the phenomena of interest, rather than the phenomena themselves. Both physical and virtual models support (elementary to undergraduate) students' understanding (Klahr, Triona, and Siler, 2008; Klahr, Triona, and Williams, 2007; Zacharia and Constantinou, 2008). However, physical objects are more advantageous in domains requiring physical manipulation and tactile senses to make effective connections (Eberbach and Crowley, 2005; Leinhardt and Crowley, 2002).

The use of models can be enormously valuable in helping students understand a process, but connecting the processes observed in a model with the full-scale phenomenon is difficult (Gentner and Colhoun, 2008; Gentner and Toupin, 1986; Jones et al., 2007a). Valuing and allowing students to talk and collaborate with one another can help overcome difficulties inherent in transferring knowledge gained through the use of models

(Dickerson and Dawkins, 2004; Johnson, 1998; Mason and Santi, 1998; Tytler, 2000; Tytler and Peterson, 2000). Students' use of words—both scientifically acceptable as well as everyday language—does not necessarily represent their understanding, and thus educators need to give students opportunities to explain, state, and clarify their thinking before drawing conclusions. Finally, a knowledgeable facilitator is critical; someone needs to be available to offer students' intellectual support and guidance as they need it.

Systems thinking and understanding emergent properties are important parts of developing an understanding of the sciences related to NOAA's mission as well as promoting stewardship of the resources that NOAA protects. Systems thinking includes being able to envision how individual components of an object or process work together to perform a function, as well as being able to simultaneously consider the effect of multiple causal factors. Emergent properties are the result of dynamic interactions among system components, and thus emergent phenomena have aggregate impacts on systems that are qualitatively distinct from the sum of the impact of individual components. Learning about complex systems can be supported by breaking ideas down into structures, behaviors, and functions (Hmelo-Silver, Marathe, and Liu 2007) and by using a multiyear learning progressions that span an entire school curriculum (Fortus et al., 2006).

In many sciences, including those that are critical to NOAA's mission, most major insights are not accomplished only through classic laboratory experimentation that most teachers have been taught (Edwards, 1997; Uthe, 2000; Windschitl, Thomson, and Braaten, 2008). Other modes of inquiry often applied in fields related to NOAA's mission include methodical observation of variations across space and changes through time and construction of computational and physical models (Kastens and Rivet, 2008; Kastens et al., 2009). Developing educational materials and programs that describe the full range of scientific modes of inquiry will support the accurate understanding of the processes of science, encouraging individuals to appropriately value scientific findings that result from a wide range of scientific practices (Kastens and Rivet, 2008; Windschitl, Thomson, and Braaten, 2008).

In summary, teaching scientific concepts related to NOAA's mission requires tools, resources, and interventions that facilitate understanding of complex systems, allow students to manipulate (physical and virtual) models that make the system framework explicit, and give them extended experiences with, discussion of, and exposure to the complex system. As education programs are developed and revised, these lessons learned should be broadly disseminated and used when appropriate.

Behavior Change and Stewardship

Many of NOAA's education efforts include stewardship goals that aim to change behaviors and attitudes. Changing behaviors and attitudes toward the environment is complex and difficult to accomplish. It is not simply the case that environmental knowledge and environmental awareness lead to changes in environmental behavior (Kollmuss and Agyeman, 2002). A review of decades of research on climate change found major psychological barriers to public engagement, including:

1. uncertainty, mistrust of, and disbelief in risk messages from scientists and government officials;
2. denial that climate change is occurring or that human activities have anything to do with it;
3. underestimation of risks resulting from climate change;
4. a sense that humans can't affect changes in the future climate; and
5. ingrained behaviors and habits (Swim et al., 2009).

Lessons from environmental education and behavior change research can inform how efforts to change behaviors are developed, implemented, and evaluated. For example, responsible environmental behaviors are associated with

- environmental sensitivity (i.e., feelings of comfort in and empathy toward natural areas);
- knowledge of ecological concepts;
- knowledge of environmental problems and issues;
- skill in identifying, analyzing, investigating, and evaluating environmental problems and solutions;
- beliefs and values regarding problems/issues and alternative solution/action strategies;
- knowledge of environmental action strategies;
- skill in using environmental action strategies; and
- belief that an individual, by working alone or with others, can influence or bring about the desired outcomes (National Research Council, 2002).

The final factor, the belief that an individual can influence the desired outcomes, is possibly the most important in promoting stewardship behaviors.

NOAA's environmental education programs need to promote these factors if they are going to lead to stewardship behaviors. Education activities with stewardship goals need to promote more than a greater understanding of the natural world: they need to also address individuals' beliefs and decision making. The environmental education guidelines of the North American Association for Environmental Education (see Box 3.3) can inform the programs with stewardship goals. NOAA can use these standards, in addition to education standards (e.g., the *National Science Education Standards*) and literacy principles they currently consider, to improve their current programs and develop new programs.

BOX 3.3
North American Association for Environmental Education Guidelines for Initial Preparation of Environmental Educators

Theme 1: Environmental Literacy
- Questioning and analysis skills
- Knowledge of environmental processes and systems
- Skills for understanding and addressing environmental issues
- Personal and civic responsibility

Theme 2: Foundations of Environmental Education
- Fundamental characteristics and goals of environmental education
- How environmental education is implemented
- Evolution of the field

Theme 3: Professional Responsibilities of the Environmental Educator
- Exemplary environmental education practice
- Emphasis on education, not advocacy
- Ongoing learning and professional development

Theme 4: Planning and Implementing Environmental Education Programs
- Knowledge of learners
- Knowledge of instructional methodologies
- Planning for instruction
- Knowledge of environmental education materials and resources
- Technologies that assist learning
- Settings for instruction
- Curriculum planning

Theme 5: Fostering Learning
- A climate for learning about and exploring the environment
- An inclusive and collaborative learning environment
- Flexible and responsive instruction

Theme 6: Assessment and Evaluation
- Learner outcomes
- Assessment that is part of instruction
- Improving instruction

SOURCE: North American Association for Environmental Education (2004).

Informal Learning Environments

Many NOAA education initiatives for both adults and children occur in informal learning environments, such as after-school or summer programs, learning centers, museums, aquariums, watersheds, and over the Internet. These environments provide direct experiential learning venues for students, teachers, and the public, as well as opportunities for instructional interaction between participants and scientists and for engaging in real scientific research activities.

Informal learning environments are particularly suitable for developing and validating learners' positive science-specific interests, skills, emotions, and identities. In fact, national surveys note that affective, emotional, and personal experiences with the ocean and climate have significant influences on adults' knowledge, attitudes, behavior, perceptions of risk, and policy preferences (Bord et al., 2001; Leiserowitz, 2006; Steel, Lovrich, et al., 2005; Steel, Smith, et al., 2005). Outdoor informal learning environments that offer personal experiences and direct contact with nature contribute to the understanding of and commitment to environmental conservation and stewardship (Bogner, 1998; Dillon et al., 2006). The need for personal experiences is not surprising, as they have long been argued to be essential to learning (Dewey, 1938).

In general, learning science in informal environments involves developing positive science-related attitudes, emotions, and identities, learning science practices, appreciating the social and historical context of science, and cognition (National Research Council, 2009). Experiences in these environments often serve as an "on ramp" to help the learner build familiarity with the natural and designed world and to establish the experience base, motivation, and knowledge that fuel and inform later science learning experiences.

The informal nature of any learning environment can vary. Learning environments defined as informal most often include learner choice, low-consequence assessment, and structures that build on the learners' motivations, culture, and competence. In designed settings (i.e., museums, science centers, aquariums, and environmental centers) experiences are typically less structured for family and peer groups than for student field trip groups. Experiences in after-school, summer, and adult programs are usually more structured than trips to designed settings with families and friends; however, they are not as structured as experiences in typical classroom settings. In general, all informal learning environments provide a safe, nonthreatening, open-ended environment for engaging with science (National Research Council, 2009).

A number of effective practices in developing informal environments for science learning can be employed by NOAA as it implements the strategic plan:

- Design with specific learning goals in mind.
- Be interactive.
- Provide multiple ways for learners to engage with concepts, practices, and phenomena in a particular setting.
- Facilitate science learning across multiple settings.
- Prompt and support participants to interpret their learning experiences in light of relevant prior knowledge, experiences, and interests.
- Support and encourage learners to extend their learning over time (National Research Council, 2009).

Teacher Professional Development

Some NOAA education programs provide time, support, and materials for teacher professional development. Teacher professional development is a continuous, lifelong process, and opportunities should connect to teachers' work in the context of the school (National Research Council, 1996). Research on effective teacher professional development can inform the continual improvement and development of NOAA's teacher professional development programs. We provide a brief overview of important lessons learned that NOAA could benefit from following.

Features of well-structured teacher development include a focus on a specific content area and teaching strategies, sustained support over time, clear connections to the classroom and the curriculum being taught, collective participation, active learning, and opportunities to practice and apply what is learned in real-world contexts (Ball and Cohen, 1999; Barnett, 1998; Garet et al., 2001; Mewborn, 2003; National Research Council, 2007; Schifter, 1996). In addition, effective professional development uses a continuous cycle of exploring new issues and problems, creating cognitive dissonance, engaging in collaborative discussions, constructing new understanding, and improving professional practice (Ball and Cohen, 1999). Professional development aimed merely at collaboration without a specific focus on topics, such as student thinking, content, or curriculum, is not as effective (Gerhart et al., 1999; Huffman, Thomas, and Lawrenz, 2003).

All teacher professional development does not have the same goals. The range of goals includes immersion, examining practice, curriculum implementation, curriculum development, and collaborative work (Loucks-Horsley et al., 1998). Professional development with any of these goals can be designed in a variety of contexts, as well as over different periods of time.

Frameworks for developing and evaluating teacher professional development exist. One framework, focused specifically on the myriad of needs and concerns associated with science education design, has a sequence of planning, input, iteration, and evaluation (Hewson, 2007; Loucks-Horsley et al., 2003). This sequence lends itself to a highly dynamic process as evaluation data are fed back into the planning loop for more effective and relevant professional development offerings. An additional framework for evaluating professional development suggests that it progresses through five levels: (1) participants' reactions, (2) participants' learning, (3) organization support and change, (4) participants' use of new knowledge and skills, and (5) student learning outcomes (Guskey, 2000). Each level is important in its own right, and each must be achieved before the next can be accomplished. The effectiveness of the professional development at each level should be ascertained.

Broadening Participation

A stated goal of many of NOAA's education initiatives is to serve a diverse set of audiences. The diverse skills and orientations that members of different cultural communities bring to formal and informal science-learning contexts are assets for NOAA to build on. For example, children reared in rural agricultural communities who have more intense and regular interactions with plants and animals develop more sophisticated understanding of ecology and biological species than urban and suburban children of the same age (Carey, 1985; Coley et al., 2005; Inagaki, 1990). In addition, there are connections between children's culturally based storytelling and argumentation and science inquiry. Two promising insights into how to better support science learning among people from nondominant backgrounds are that learning environments should be developed and implemented with the interests and concerns of community and cultural groups in mind (e.g., project goals should be mutually determined by educators and the communities and cultural groups they serve) and in ways that expressly draw on participants' cultural practices, including everyday language, linguistic practices, and common cultural experiences (National Research Council, 2009).

Community-based programs that involve diverse learners in locally defined science inquiry, such as identifying and studying local health and environmental concerns, show promise for developing sustained, meaningful engagement. Specific cultural resources can also be harnessed in programs designed to support the development of materials aligned to the needs, interests, and knowledge of the target audience. In addition, to serve the goal of broadening participation in science, front-line staff should have the disposition and repertoire of practices and tools at their disposal to help learners expand on their everyday knowledge and skill to learn science. In order to accomplish this, practitioners need professional development to support their efforts.

Partnerships between science-rich institutions such as NOAA and local communities show great promise for fostering inclusive science learning (National Research Council, 2009). Developing productive partnerships requires considerable time and energy. Effective strategies for organizing partnerships include identifying shared goals; designing experiences around issues of local relevance; supporting participants' patterns of participation (e.g., family structure, modes of discourse); and designing experiences that satisfy the values and norms and reflect the practices of all partners (National Research Council, 2009). More details on establishing a productive partnership are provided in Chapter 4.

Workforce Development Programs

Developing Accurate Workforce Needs Assessments

As a first step in addressing the workforce goals, NOAA needs a sense of the current and future workforce needs related to its mission. In general, the discussion of U.S. STEM workforce needs can be divided into two broad categories: higher level studies of the overall issue, such as *Rising Above the Gathering Storm* (National Academy of Sciences, National Academy of Engineering, and Institute of Medicine, 2006), and more focused industry or area-specific studies. For example, the Marine Advanced Technology Center is currently undertaking a study of the ocean science, technology, and operations workforce (Sullivan, Rosenfeld, and Murphree, 2007). To date the only workforce studies that focus on the fields related to NOAA's mission are two congressionally mandated studies of the fishery stock assessment workforce (U.S. Department of Commerce and U.S. Department of Education, 2008) and a study of the geoscience workforce (American Geological Institute, 2009). The fishery stock workforce assessment is limited in its focus to only one job in the fishery workforce—fishery stock assessment. Some of the other areas in which fisheries also need experts are conservation biology, environmental conservation, fish ecology, and fishery operations and management. Also, the geoscience workforce report only summarizes the state of the current workforce, stopping short of estimating current or future workforce needs in any fields or estimating whether there are sufficient students in the workforce pipeline to fill future workforce needs. Without clear estimates of workforce need and workforce preparation, federal agencies will be unable to gauge how to focus workforce development initiatives.

The lack of workforce studies that focus on the specific industries and fields related to NOAA's mission is not surprising given the complexity and cost of such efforts. First, data on employment demand are difficult to obtain, particularly broken down by relevant skill areas, and those data and projections that exist are often ambiguous if one looks beyond the near-term future (National Research Council, 2006). In addition, workforce needs assessment is often misconstrued as a simple task of comparing graduation rates in particular fields with the occupational needs of a related field. Two factors lead to a disconnect between graduation trends and occupational trends. First, people with particular education specializations and skills can often use them in cross-fertilizing other fields. For example, in 2003 over 75 percent of individuals with a degree in aerospace engineering or space science were not employed as aerospace engineers or space scientists (National Research Council, 2006). Second, nearly all fields require a workforce with a wide range of educational backgrounds and expertise. For example, the ongoing study of the ocean-observing systems workforce will assess workforce needs across operation and maintenance of facilities positions; platforms and instrumentation positions; the data and information management positions; and education outreach and applications positions (Sullivan et al., 2007).

Critical Factors in Supporting Workforce Development

The majority of NOAA workforce development programs consist of support of undergraduate, graduate, and postdoctoral students. Important factors in higher education programs that NOAA can employ to address workforce needs are research experience, mentoring, career development, and funding. Research experiences that include laboratory and field experiences, research seminars and workshops, and opportunities to make research presentations and teach science have been cited as critical to undergraduate, graduate, and postdoctoral students (Gilligan et al., 2007; National Academy of Sciences, National Academy of Engineering, and Institute of Medicine, 2000; National Research Council, 2005). These opportunities are also important for postdoctoral students because they provide opportunity to choose a subspecialty and develop research independence and bring their knowledge and expertise to a range of careers.

Many of NOAA's education programs that aim at broadening participation in the fields critical to the agency's mission support students at minority-serving institutions. These institutions tend to focus on providing access to higher education for lower income and minority students, and tend to focus on teaching rather than research (National Research Council, 2005). These institutions tend not to have the kind of research infrastructure that a more research-intensive institution provides. Students at these institutions can benefit from partnerships with research-intensive institutions in various ways, including mentorship by research scientists, postdoctoral associates and graduate students from research-intensive universities, shipboard research experiences on the research-intensive universities' research vessels, and increased opportunities to participate in research (Gilligan et al., 2007).

Another important factor in higher education programs is funding (National Research Council, 2005). Funding for undergraduate and graduate students can come from stipends, research positions, scholarships, tuition support, and grants. Providing funding to students creates "protected time" that allows them to focus on their research and classroom responsibilities. This is critical because undergraduate and graduate students often must take on additional outside work in order to make ends meet. This has been cited as a recipe for disaster for lower income students, because it constitutes a barrier against participation in research programs.

Mentoring has also been cited as a valuable feature of workforce development programs (National Research Council, 2005). Productive mentoring provides guidance in four key areas: (1) improving the trainee's research skills, (2) providing motivation and personal growth, (3) providing career guidance, and (4) promoting the trainee for scholarships and other development opportunities. Negative mentoring experiences are associated with being given mundane administrative tasks to perform in lieu of experiments, being ignored by the mentor, and not receiving encouragement. Training in science fields historically assumes that if one is trained, one will therefore be a good trainer (mentor), but this is not necessarily so (National Research Council, 2005). Mentoring is a skill for which academic researchers rarely receive any formal training. Thus, workforce development programs that support higher education research laboratory and field experiences should include training for student mentors.

In addition, career development opportunities are an important aspect of undergraduate and graduate education programs (National Research Council, 2005). Programs should provide undergraduate and graduate trainees with opportunities to network and collaborate with other scientists. Making these connections influences decisions to pursue graduate education and the fields in which employment or postdoctoral experiences are sought.

At the postdoctoral level, three guiding principles for effective programs have been recommended (National Academy of Sciences, National Academy of Engineering, and Institute of Medicine, 2000). First, postdoctoral experience is first and foremost a period of apprenticeship for the purpose of gaining scientific, technical, and professional skills that advance one's professional career. Second, postdoctoral students should receive appropriate recognition (including lead author credit) and compensation (including health insurance and other fringe benefits) for the contributions they make to the research enterprise. Third, to ensure that postdoctoral appointments are beneficial to all concerned, all parties to the appointments—the student, the adviser, the host institution, and funding organizations—should have a clear and mutually agreed-on understanding with regard to the nature and purpose of the appointment. Actions recommended for creating effective postdoctoral programs by the National Academy of Sciences, National Academy of Engineering, and Institute of Medicine (2000) appear in Box 3.4.

BOX 3.4
Guidance for Developing Successful Postdoctoral Programs

In order to enhance the postdoctoral experience, advisers, institutions, funding organizations, and disciplinary societies should

1. Award institutional recognition, status, and compensation commensurate with the contributions of postdocs to the research enterprise.
2. Develop distinct policies and standards for postdocs, modeled on those available for graduate students and faculty.
3. Develop mechanisms for frequent and regular communication between postdocs and their advisers, institutions, funding organizations, and disciplinary societies.
4. Monitor and provide formal evaluations (at least annually) of the performance of postdocs.
5. Ensure that all postdocs have access to health insurance, regardless of funding source, and to institutional services.
6. Set limits for total time of a postdoc appointment (of approximately five years, summing time at all institutions), with clearly described exceptions as appropriate.
7. Invite the participation of postdocs when creating standards, definitions, and conditions for appointments.
8. Provide substantive career guidance to improve postdocs' ability to prepare for regular employment.
9. Improve the quality of data both for postdoctoral working conditions and for the population of postdocs in relation to employment prospects in research.
10. Take steps to improve the transition of postdocs to regular career positions.

SOURCE: National Academy of Sciences, National Academy of Engineering, and Institute of Medicine (2000).

4

Overview and Critique of NOAA's Education Programs

The National Oceanic and Atmospheric Administration (NOAA) supports a wide array of education activities and products that serve K-12 and higher education students as well as the public. The agency's activities and products range from hands-on learning experiences, to teacher training initiatives, to a spherical data visualization tool, to literacy frameworks. They are implemented in formal and informal learning environments and address the range of scientific and stewardship issues at the heart of the agency's mission. NOAA has created these activities and products in partnership with other federal agencies, nonprofit organizations, industry, state, and local education agencies, and other groups. In this chapter we provide an overview of how the education programs are organized, describe and critique the education programs, and discuss cross-cutting issues, including addressing core science and engineering principles, the quality of NOAA education websites, and portfolio balance.

Table 4.1 lists the education programs supported by five line offices and the Office of Education. Some programs are the result of the efforts of more than one office. Also, education programs are being carried out at multiple levels: that is, education is supported by line office managers, as well as by offices within the line offices, such as the National Sea Grant College Program Office and the Office of National Marine Sanctuaries. The majority of NOAA staff, offices, and sites are located in a coastal region (see Figure 4.1), and most education programs are supported or run by education staff at these offices and sites (a few programs are national in scope and not tied to a specific region). Thus, the majority of the education initiatives serve students, teachers, and the public residing in coastal regions.

TABLE 4.1 Education Programs and Supporting Offices

Education Program	Supporting Office
Ocean Exploration and Research's Ocean Hall	Oceanic and Atmospheric Research
Climate Program	Oceanic and Atmospheric Research
Sea Grant	Oceanic and Atmospheric Research
Ocean Exploration and Research's Ocean Explorer	Oceanic and Atmospheric Research National Ocean Service
Coral Reef Conservation Program	Oceanic and Atmospheric Research National Marine Fisheries Service National Environmental Satellite, Data, & Information Services National Ocean Service
Teacher at Sea	National Marine Fisheries Service National Ocean Service's National Marine Sanctuaries
Storm/Tsunami Ready	National Weather Service
Cooperative Institutes	Oceanic and Atmospheric Research National Marine Fisheries Service National Weather Service National Environmental Satellite, Data, & Information Services
National Marine Sanctuaries	National Ocean Service
National Estuarine Research Reserve System	National Ocean Service
Environmental Literacy Grants	Office of Education
Bay-Watershed Education and Training	Office of Education National Ocean Service's National Marine Sanctuaries and National Estuarine Research Reserve System National Ocean Service's Coastal Service Center
Educational Partnership Program	Office of Education
Dr. Nancy Foster Scholarship Program	Office of Education
JASON Project	Office of Education
Ernest F. Hollings Scholarship Program	Office of Education

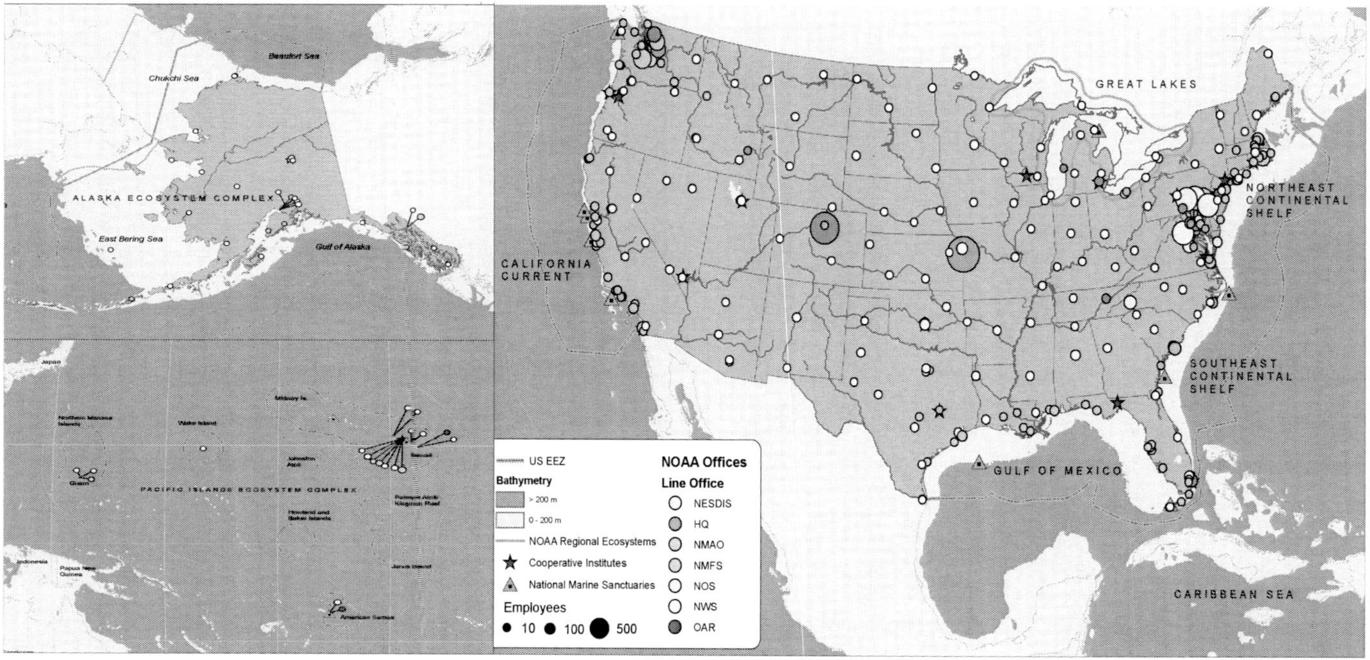

FIGURE 4.1 NOAA offices and sites around the country.
NOTES: CRCP = Coral Reef Conservation Program; ELG = Environmental Literacy Grants; EPP = Educational Partnership Program; NERRS = National Estuarine Research Reserve System; NESDIS = National Environmental Satellite, Data, and Information Service; NMFS = National Marine Fishery Service; NWS = National Weather Service.
SOURCE: National Oceanic and Atmospheric Administration.

Table 4.2 summarizes the education activities of each program, although there is some ambiguity in determining which programs conduct which type of activities. Some programs run many different initiatives, and others focus significant effort on just a few. There is large variation in the number of initiatives and amount of funding across programs. For example, both the National Sea Grant College Program and the National Marine Fisheries Service conduct various types of education activities and have multiple local sites. Yet the Sea Grant Program is very large and has produced large amounts of curricular materials, conducts several large fellowship initiatives,[1] and supports many outreach initiatives. In contrast, the Fisheries Service has a very small education budget that supports one curriculum, one fellowship program, the Teacher at Sea Program, and several small outreach initiatives. Thus, the differences between these programs cannot be captured by a yes or no. It is likely that NOAA is having trouble categorizing the education activities under each line and program office. A clearer table that organizes education programs by focus, goals, and scope would be a useful piece of the implementation plan that is currently under development.

The programs could be grouped in various ways, and we chose to group them into four main categories by focus: (1) place-based education that supports local education efforts, (2) place-based education that occurs at NOAA managed sites, (3) nationwide curricula and teacher professional development, and (4) higher education.

COMMITTEE REVIEW

Information was collected on each program through presentations to the committee, phone interviews and e-mails with the education staff, site visits, and review of online materials. Further details on most programs can be found at their websites (listed in Appendix C).

Place-Based Education That Supports Local Education Efforts

Bay-Watershed Education and Training

The Bay-Watershed Education and Training (B-WET) Program is an environmental education effort promoting locally relevant, experiential learning for K-12 students and teachers. The program aims to develop a well-informed citizenry involved in decision making that positively impacts coastal, marine, and watershed ecosystems. B-WET is based on the idea that firsthand experience, in the context of one's community and culture,

[1] A complete list of fellowship, scholarship, and internship opportunities at NOAA is found at http://www.oesd.noaa.gov/noaa_student_opps.html [accessed May 2010].

TABLE 4.2 Education Program Activities

Program	Grants for External Education	NOAA On-Site Education	Web-Based Education	Locally-Based	Curriculum Development	Teacher Professional Development	Fellowships for Higher Education	Informal Education Outreach
B-WET	•			•		•		•
Coral Reef Conservation	•		•	•	•	•		•
Storm/Tsunami Ready			•	•				•
Estuary Research Reserves		•	•	•	•	•	•	•
Fisheries Service		•	•	•	•	•	•	•
Marine Sanctuaries		•	•	•	•	•		•
Environmental Literacy Grants	•				•	•		•
JASON			•		•	•		•
Ocean Explorer	•		•		•	•		•
Teacher at Sea/in Air						•		
Cooperative Institutes	•			•	•	•	•	•
Educational Partnerships	•			•			•	•
Hollings							•	
Foster							•	
Sea Grant	•		•	•	•	•	•	•
Climate Change Program			•		•	•	•	•
Ocean Hall			•					•
National Weather Service			•			•		•
National Ocean Service			•		•	•		

NOTE: A circle in the box indicates that the program has at least one initiative in an area.

B-WET Quick Facts
Scientific Focus—Watershed ecosystems and conservation
Audience—Primarily K-12 students and teachers
Format—Grants primarily for education activities and professional development
Education Budget—$4.2M in fiscal year (FY) 2007
Supporting Office—Office of Education

is an important element in fostering environmental stewardship. B-WET is an Office of Education program that has been administered by regional program offices (e.g., the Office of National Marine Sanctuaries, the NOAA Chesapeake Bay Office, and the Pacific Services Center) since 2002. The program expanded to three additional regional offices in 2008 (Gulf Coast, New England, and Pacific Northwest).

B-WET provides competitive funding ($4.2 million in 2007) from regional program offices to local grantees that promote what it calls Meaningful Watershed Educational Experiences. Such an experience "integrates field experiences in a watershed with multidisciplinary classroom activities and instruction. Students then share their discoveries about the watershed with local schools and communities, both orally and in written form" (National Oceanic and Atmospheric Administration, 2009). To date 353 grants have been funded by B-WET for a total of $20 million. The grants typically provide partial funding for education activities, with matching or additional funding coming from other agencies or private-sector organizations. These grants have supported programs that reached about 18,000 teachers, 1,000 other adults, and 160,000 students. Grant recipients include K-12 public and independent schools and school systems, institutions of higher education, commercial and nonprofit organizations, community organizations, and state and local governments. B-WET is currently implementing a programwide evaluation plan that will provide a common framework for evaluation across regions.

Impact: Evaluations of local B-WET activities, which included control groups, have shown the program to be successful in increasing teachers' confidence in their ability and intentions to implement Meaningful Watershed Educational Experiences. Students gained knowledge about environmental stewardship and watershed and marine sanctuaries and seemed to enjoy learning about and protecting the ocean. Program-supported activities also increased student and teacher understanding about preventing pollution.

Coral Reef Conservation Program

The Coral Reef Conservation Program (CRCP), established by the 2000 Coral Reef Conservation Act, is a partnership between the National Ocean Service, the National Environmental Satellite Data and Information Service, the National Marine Fisheries Service, and the Office of Oceanic and Atmospheric Research. It brings together expertise from across NOAA for a multidisciplinary approach to understanding coral reef ecosystems and supporting education activities. The program facilitates and supports partnerships with scientific, private, government, and nongovernment groups at local, state, federal, and international levels.

The education strategy seeks to deliver two messages. First, coral reefs are valuable resources. Second, the health of coral reef ecosystems is at serious risk from a variety of human activities. The goals of the education program are to raise public awareness and appreciation for coral reef ecosystems, incorporate coral reef issues in education programs, increase assessments and monitoring of coral reef habitats, and support local education initiatives through grants. The program has developed a suite of educational and professional development resources for teachers wanting to explore coral reef ecosystem science in their classrooms. These include online discovery kits, an educational resources CD, professional development workshops, and a coral reef and satellites curriculum. There have been several local CRCP outreach activities, including media outreach, environmental expos, and other communication efforts.

CRCP Quick Facts
Scientific Focus—Coral reef ecosystems and conservation
Audience—General public, K-12 educators
Format—Varied, grants for local initiatives, some curriculum development and teacher professional development
Education Budget—$.9M in FY2007
Supporting Offices—Oceanic and Atmospheric Research, National Marine Fisheries Service, National Environmental Satellite, Data, & Information Services, National Ocean Service

Impact: The committee did not have access to assessments of local CRCP education initiatives, and no national education products have been evaluated. It is therefore challenging to determine how effective

CRCP has been in reaching its educational goals. It is clear that program leveraged funds and partnerships to enhance local efforts in coral reef conservation outreach. A seven-member expert external panel review commissioned by NOAA to inform the 2009-2011 CRCP strategic plan made one recommendation related to education: the CRCP's potential would be more likely to be fulfilled if it provided "better general education and outreach on the threats to coral reefs, the likely social and economic consequences of their loss, and the measures that can be taken to ensure their continued survival and productivity. NOAA's role in this should be clearly defined and should include use of existing and development of new products, provision of expert information, and development of partnerships explicitly for communication" (Coral Reef Conservation Program, 2007, pp. 5-6).

National Weather Service

The National Weather Service (NWS) provides a variety of education resources to classrooms and the public, although it does not have a concerted education program. Most activities are informal, like the NOAA/American Meteorological Service WeatherFest, an interactive science festival for the public. The festival includes a teacher training workshop as well as hands-on activities run by teachers and scientists on weather and climate topics, drawing approximately 3,000 visitors per year. NWS has supported formal education activities, such as the creation of the Xtreme Weather CD. It was created by the Illinois Education Association and the NWS as one of several partners on the project. In addition, NWS field office staff make approximately 2,400 school visits each year.

NWS Quick Facts
Scientific Focus—Weather and meteorology
Audience—K-12 students and teachers
Format—classroom visits, informal activities, and online education materials
Education Budget—$3.4M in FY2007
Line Office—National Weather Service

Since 1999, the NWS has supported StormReady and TsunamiReady—nationwide community preparedness programs that use a grassroots approach to help communities develop plans to handle severe weather (e.g., thunderstorms, tornados, hurricanes) and tsunamis. StormReady and TsunamiReady help community leaders and emergency managers strengthen their local hazardous weather operations, supplying them with targeted publications about severe weather safety, related statistics, relevant presentations, and brochures and requiring that community seminars are held to promote the importance of public readiness.

Impact: Overall, NWS is responsive to requests for visits by schools and teachers and has carried out a large number of education activities in relation to the level of staff and budget available for education purposes. This results in sporadic use of resources. In the absence of specific evaluations, a first-principles approach suggests that such efforts would be more efficient in terms of human and financial resources if the staff at NWS field offices were better coordinated. As of February 2009, there were 1,411 StormReady sites in 50 states, Puerto Rico, and Guam and 63 TsunamiReady sites in 10 states, Puerto Rico, and Guam.

TsunamiReady is a relatively new program with a more robust educational and outreach presence than StormReady. StormReady is focused on providing informational resources but does not provide many education activities for communities and does not seem to be reaching individuals who are not already interested in weather, especially those from underserved populations. The program could increase its reach by increasing its engagement with underserved communities.

Place-Based Education That Occurs Primarily at NOAA Sites

National Estuarine Research Reserve System

The National Estuarine Research Reserve System (NERRS) is a network of 27 protected areas in the National Ocean Service, established by the 1972 Coastal Zone Management Act to promote long-term research, education, and stewardship. This program requires partnerships between NOAA and the coastal states where the estuaries exist. NOAA provides partial funding (70 percent from NOAA, 30 percent from state agencies), national guidance, and technical assistance. Each reserve is managed on a

NERRS Quick Facts
Scientific Focus—Estuary ecosystems and conservation
Audience—Primarily K-12 students and teachers
Format—Varied, many informal programs plus some curriculum, professional development, and graduate fellowships
Education Budget—$1.1M in FY2007
Line Office—National Ocean Service

daily basis by a lead state agency, nonprofit organization, or university with input from local partners. The reserves take a local approach to national priorities. Land use and population growth, water quality degradation, habitat loss and alteration, and changes in biological communities are the core topics for science education and training.

The goals of NERRS education activities are to enhance public awareness and understanding of estuarine areas and provide suitable opportunities for public education and interpretation. With the support of NOAA, state agencies, and other partners, the reserves run approximately 3,000 K-12 programs. Approximately 3,500 teachers have participated in reserve professional development, and 700,000 people have watched an EstuaryLive program (the NERRS online education tool, which includes teacher professional development resources, an estuary curriculum, and classroom activities). Graduate students are served by the Graduate Research Fellowship Program, which awards stipends to 54 master's and doctoral students per year. Other initiatives are locally based and include professional teacher training programs, K-12 student programs, and community outreach and educational programs.

Impact: A 2003 review of NERRS education found that there had been little formal evaluation of the K-12 and teacher professional development programs and that local reserves have focused their evaluations on the number of people participating rather than changes in their knowledge, intentions, or actions (Pandion Systems, 2003). Since 2003, NERRS has improved its evaluation strategy and now implements performance measurements and more frequently implements pre-post participant surveys, site-based evaluations, evaluations of national programs (i.e., EstuaryLive), and program reviews (the most recent education program review report was released in 2003). Through these mechanisms NERRS has become better at measuring participants' knowledge, intentions, and actions.

It is important that NERRS continue with this enhanced evaluation strategy for its various information reporting requirements and needs. EstuaryLive has been evaluated and shown to be effective at holding student interest and increasing their estuary content knowledge (Pandion Systems, 2005, 2006). Surveys did not attempt to measure excitement for science, intentions, or behavioral changes, which might have better captured the stewardship potential of the program. With the assistance of a nonprofit educational research and development firm, NERRS developed the Estuaries 101 curriculum based on the needs of teachers and schools, and aligned it with big ideas in science and ocean and climate literacy (Hammerman, 2007). Organizationally, NERRS is extremely decentralized, but staff from each site meet yearly. Overall, the program has made changes to improve evaluation practices, and positive impacts of the program have been documented.

Office of National Marine Sanctuaries

The Marine Protection, Research and Sanctuaries Act established the Office of National Marine Sanctuaries (ONMS) Program in 1972. The sanctuaries are responsible for promoting public understanding of national marine sanctuaries, national marine heritage, and the marine environment. The National Marine Sanctuaries Act explicitly mandates that the sanctuaries support education and outreach activities. Today, 13 marine sanctuaries and the Papahānaumokuākea Marine National Monument promote public understanding of marine sanctuaries and the marine environment.

Each site has outreach and education functions related to common themes, which are generally handled by one or more education coordinators or specialists. The national office also supports large- and small-scale national education activities, including OceansLive and Data in the Classroom. Educational materials for students and teachers are provided online, and hands-on education experiences are available at each sanctuary. A wide range of formal and informal education and outreach activities are supported, including naturalist volunteer programs, adult education, museum exhibits, community college courses, teacher professional development, summer camps, ocean literacy partnerships, student field days, art contests, and curriculum. In addition, each site has either a visitor center or multiple permanent sanctuary exhibits displayed at partner facilities. An internal mini-grants program, led by the national education coordinator, encourages local sanctuaries to collaborate with each other and develop new educational and outreach programs. This coordinator also leads an Education Executive Council, comprised of a site representative from each of the four regions, which works on systemwide education policy issues.

Impact: ONMS has been proactive in increasing the quality of its education evaluations. It is encouraging that it is working to identify objectives and student outcomes for all of its education initiatives. It is especially important that the office is studying how current evaluations, which fall short of measuring these outcomes, can be improved. While this work is ongoing, a complete program evaluation is not available. Existing evaluations, however, are very positive. Several teacher professional development

ONMS Quick Facts
Scientific Focus—Marine sanctuary ecology and conservation
Audience—K-12 students and teachers, general public
Format—Varied, informal and formal
Education Budget—$2M in FY2007
Line Office—National Ocean Service

workshops, including Dive into Education (Office of National Marine Sanctuaries, 2004), the LiMPETS Workshop (Office of National Marine Sanctuaries, 2005a), and the Hawaii Field Study Workshop (Office of National Marine Sanctuaries, 2005b), were evaluated and found to be highly successful. Teachers ranked the workshops as among the best they had attended, stating that they were more likely to teach ocean science after attending. The Hawaii Field Study was successful in promoting teaching and behavioral changes among participants (Office of National Marine Sanctuaries, 2005b).

National Marine Fisheries Service

The National Marine Fisheries Service (NMFS), founded in 1871, is responsible for the management, conservation, and protection of living marine resources and their habitat within the United States' Exclusive Economic Zone (waters 3-200 miles offshore). The office supports a large number of extension programs and outreach for professionals in the fishing industry as well as policy makers.

The NMFS education program was established in 2008,[2] and recently the position of national education coordinator was created to coordinate efforts across the science centers and to develop national programs. NMFS outreach and education activities aim to align their goals with NOAA's overall strategic education plan. Although it has limited budget and staff dedicated to education and outreach, a variety of education initiatives are carried out by regional science centers, including internship programs, workshops and science festivals, presentations to community members or students, teacher professional development, and curriculum development. Many of the initiatives are informal and combine outreach and education. NMFS makes extensive use of partnerships with schools, other NOAA offices, aquariums, and other community and environmental groups. In addition, many NMFS research grants include funding for undergraduate and graduate students.

Impact: There are no evaluations of NMFS education activities. The activities run by the NMFS are highly decentralized, and there is a great diversity in programming among the centers. Some of the variation stems from the amount of staff dedicated to education at each center, which ranges from a scientist authorized to spend less than 30 percent of his or her time on education to one full-time staff position. Almost all education initiatives are run by scientists with little education background or training. It is commendable that so many scientists volunteer their time to participate in the education initiatives, yet having an effective education program also requires a dedicated education

[2]In FY2008 NMFS received funding specifically for education. Information regarding the amount of education funding received was not available early enough to be included in this report.

NMFS Quick Facts
Scientific Focus—Ocean and environmental science
Audience—Varied
Format—Varied, mostly informal
Education Budget—$1.1M in FY2007 for outreach, no education-specific funding
Line Office—National Marine Fisheries Service

staff. The new national education coordinator can be a first step in producing a cohesive education program with clear goals and outcomes.

Nationwide Curriculum Development or Teacher Professional Development Programs

Environmental Literacy Grants

Environmental Literacy Grants (ELG) from the Office of Education provide funding for K-12 environmental literacy education initiatives, both formal and informal. The grants aim to fund initiatives that will catalyze change in K-12 education through development of new programs and materials or by supporting transformative methods. Approximately five to seven new grants are made per year totaling $4 million for FY2009 and FY2010. Individual grantees can receive $200,000-750,000. Institutions of higher education, for-profit and nonprofit organizations, state, local, and Indian tribal governments, K-12 public and independent schools, science centers, and museums are eligible to receive grants.

Initiatives are encouraged to incorporate NOAA data, data visualizations, and resources and to partner with NOAA entities or connect to previously funded grantees. In general, ELG initiatives have focused on the education of preservice teachers or the professional development of in-service teachers, and the development of K-12 curricula and related instructional materials. Priorities of the program have changed since its inception in 2005. ELG has been used to expand the reach of the Environmental Literacy Framework and to augment NOAA's research and technological advances for education efforts, including Science on a Sphere (see Box 4.1 for information on how this technology originally developed as a research tool is now used as an innovative education technology).

Impact: ELG is a relatively new program, and there is currently no programwide evaluation. The Office of Education is currently working to determine programwide metrics that can be used to compare outcomes across initiatives. Each initiative receiving grant funds is required to conduct

BOX 4.1
Science On a Sphere

Science On a Sphere (SOS), a large globe-shaped visualization system that uses computers and video projectors to display animated data onto the outside of a sphere, was invented by Alexander MacDonald, director of the NOAA Earth System Research Laboratory in Boulder, Colorado. He came up with the concept in 1995 as an outgrowth of other visualization initiatives he was directing in the predecessor Forecast Systems Laboratory. An early prototype was built in 1995 and a patent was awarded to NOAA for SOS in 2005. It can display dynamic, animated images of the atmosphere, oceans, and land of a planet. While SOS was originally developed as a research tool, NOAA has adapted it as an education and outreach tool to describe the environmental processes of the Earth and increase earth science literacy among museum and science center visitors.

NOAA has invested money primarily through the Environmental Literacy Grants program to expand the use of SOS to 28 museums and science centers. SOS is used in a wide variety of exhibit formats. Auto-play mode is the most popular format, which many institutions augment with docents to help deliver content. There is also some work being done to allow visitors to interactively control the content shown on the sphere, including selecting and rotating content. The Orlando Science Center will feature SOS as part of their upcoming Global Decision Room exhibit, which will have visitors make decisions on behalf of the global population and then see the results of their decisions played out on the sphere. While many exhibits use SOS primarily to highlight global-scale phenomena, such as in the Sant Ocean Hall, some exhibits feature content designed to highlight local phenomena. For example, the Bishop Museum has a program that uses SOS to explain why Hawaii has such a favorable climate.

The ability to insert graphic media into three windows on the sphere allows for a powerful connectivity from global scale phenomena to regional and local phenomena that can also be captured in specific exhibits outside the SOS. For example, the Ocean Explorium at the New Bedford (Massachusetts) Seaport uses SOS to highlight global issues of potential and real ocean acidification impacts on coral reefs and shellfish as a result of modification of ocean carbonate chemistry and connects this phenomenon with exhibits of living coral reefs and shellfish.

SOS has been received very positively. In a formative evaluation at the Maryland Science Center, visitors almost unanimously (98 percent) rated SOS as excellent or very good. A series of evaluations of the entire SOS program are being conducted in partnership with the Institute for Learning Innovation, the first being "Nature and Range of Impact," scheduled for completion in late 2009.

ELG Quick Facts
Scientific Focus—Environmental and earth science
Audience—Formal and informal educational institutions
Format—Grants
Education Budget—$5.5M in FY2007
Line Office—Office of Education

its own evaluation, but implementation varies widely, from reporting the numbers of participants, to measures of student knowledge, to achievement over an academic year. Many individual initiatives have been very successful at increasing interest in science among participants and increasing teacher knowledge about and ability to teach environmental science. Without a programwide evaluation, however, it is impossible to determine whether ELG is efficiently allocating money to the most successful initiatives. The ELG proposal rating rubric is designed to rate proposals that include the use of NOAA scientific data, findings, and expertise higher than those that plan to use scientific data, findings, and expertise from other sources. This practice unnecessarily limits the range of scientific data, findings, and expertise that grantees can use. The demand for NOAA expertise and data does not exist in all of its education grant programs. The committee is concerned that the emphasis on internal NOAA expertise and data ignores much valuable knowledge and information from the wider scientific and engineering community.

Office of Ocean Exploration and Research

The Office of Ocean Exploration and Research (OER), founded in 2001, is part of the office of Oceanic and Atmospheric Research (OAR). Called Ocean Explorer, it coordinates NOAA's exploration efforts and facilitates research expeditions, committing at least 10 percent of its annual operating budget to education and outreach. In partnership with the National Ocean Service, OER directs and maintains the official website for these explorations, which serves as a public archive of the exploration program, chronicling many of the missions with detailed daily logs, background essays, multimedia offerings, and over 130 hands-on lesson plans and a curriculum based on the explorations. In addition, OER supports the NOAA ship *Okeanos Explorer*, which travels around the globe to map the seafloor and characterize largely unknown areas of the ocean, supporting education activities through use of real-time broadband satellite communications to connect the ship and its discoveries live with audiences ashore.

OER Quick Facts
Scientific Focus—Deep ocean science
Audience—5-12 students and teachers, website is for general public
Format—Curriculum, professional development, and web-based media
Education Budget—$.9M in FY2007
Line Office—Office of Oceanic and Atmospheric Research

The OER education activities are intended to highlight the value of ocean exploration discovery and research in enhancing awareness, understanding, and stewardship of the intricate ocean system and its importance to life on Earth. The Ocean Explorer website offers near real-time access to a series of multidisciplinary ocean explorations through imagery, video, webcast and web chats, and topical essays. The site also includes lesson plans, curricula, and professional development opportunities. OER uses partnerships with local aquariums and science centers to provide one-day workshops for teachers and other educators on the use of the Ocean Explorer curriculum and to introduce participants to ocean scientists and explorers and their research. OceanAGE Careers allows users to interact with explorers and scientists who study the ocean world through live interviews, profiles, and mission logs.

Impact: The Ocean Explorer professional development institutes were recently evaluated by an outside evaluator (Day-Miller and Payne, 2009). Participants were very positive about their experience at the institutes and afterward 68 percent reported using the Ocean Explorer curriculum in their classroom, all of whom reported that their students enjoyed the curriculum. Participants also indicated increased enthusiasm for teaching ocean science. Scientists who participated in Ocean Exploration signature expeditions and teacher professional development institutes were also positive about their experience (Lovelace, 2006); however, no information on the impact of the scientists' contribution was assessed. It is encouraging that OER has sought to link together its many lesson plans into a coherent curriculum, yet it is unclear what aspects of environmental literacy the curriculum was designed to support or what impact it has had on student learning or teacher practices.

Teacher at Sea

The mission of the Teacher at Sea Program, managed by NMFS,[3] is to give teachers clearer insight into the ocean and a greater understanding of

[3]The management of the Teacher at Sea Program was recently transferred to NMFS. Previously it was managed by the Office of Marine and Aviations Operations.

maritime work and studies and to increase their level of environmental literacy by fostering an interdisciplinary research experience. In addition, scientists, NOAA Corps officers, and crew gain motivated volunteers to help carry out their initiatives. Since 1990 the program has provided opportunities for kindergarten through college-level teachers to participate in research at sea aboard research and survey ships under the tutelage of NOAA scientists and crew. An offshoot, the Teacher in the Air program, was first piloted in 2004. It has enabled teacher participants to observe research activities and interact with scientists while on board NOAA aircraft.

After their experience at sea or in the air, participants create a lesson plan that addresses the science and research and a lesson plan, activity, or similar document (e.g., brochure, flyer) that addresses ocean careers. After teaching their newly developed lesson plan, teachers must create and administer a test that measures students' knowledge of the lesson's subject matter as appropriate to grade level. Finally, teachers prepare an article for publication or conduct a presentation about the mission at an educators' conference or for colleagues. Depending on funding, 20-30 educators are chosen through a competitive grant process to participate each year. Over 525 teachers have participated in Teacher at Sea from 48 states, American Samoa, Argentina, Chile, and Puerto Rico.

Impact: Teacher at Sea is currently undergoing outside evaluation. There is considerable anecdotal evidence that it provides a key experience for teachers that enhances their excitement for teaching science and provides them with knowledge of scientific careers. While the high cost of the program per teacher presents a question as to its scalability, not all programs need to be scalable. It is valuable for NOAA to support some programs that have deep, long-lasting impact with a smaller audience. Currently lesson plans produced by participants are used only by the individual participants with their students. Collecting the lesson plans and making them available to the wider public might increase the reach and scope of the program without affecting the impact of the program on the participating teachers.

Teacher at Sea Quick Facts
Scientific Focus—Ocean science
Audience—K-12 and undergraduate teachers
Format—Teacher professional development
Education Budget—$.2M in FY2007
Line Office—National Marine Fisheries Service

BOX 4.2
Initiatives Focused on Diversity

B-WET-California provided guidance to all project coordinators on adapting programs to multicultural audiences at its recent annual conference. One of its initiatives that focuses specifically on multicultural audiences is Project Watershed by Reaching Out to Communities and Kids with Science in San Francisco. Project Watershed brings together 15 urban students in grades 9-12 to design and implement a year-long study of watershed water quality on San Francisco's east side. The program is designed to provide high school students interested in earth and environmental science with a long-term research study experience. Students learn all required skills to participate in the program, collect and analyze water samples, complete analysis of data, and discuss the implications. Students present their results with a poster at a major scientific conference.

Environmental Literacy Grants fund the Multicultural Students at Sea Together (MAST) Program, which is a multidisciplinary program that engages students in NOAA-related marine research and explores marine policy, the heritage of African Americans and Native Americans in the coastal environment, and seamanship. MAST students use the Chesapeake Bay to understand efforts to protect, restore, and manage the use of coastal and ocean resources through an ecosystem approach to management. Hampton University, which conceived of and runs the program, partners with NOAA, university labs, the Environmental Protection Agency, museums, the Chesapeake Bay Foundation, and the Menhaden fishing industry to expose students to diverse aspects of marine research and policy.

National Marine Sanctuary Program and B-WET are working to broaden participation to diverse groups through the Multicultural Education for Resource Issues Threatening Oceans (MERITO) Program. MERITO is an emerging multicultural education and outreach initiative aiming to build and engage a conscious and culturally inclusive constituency for ocean protection nationwide. Currently it operates at Monterey Bay and Channel Islands National Marine Sanctuaries, offering adult

Programs Primarily Focused on Higher Education

Educational Partnership Program

The Educational Partnership Program (EPP), established by the Office of Education in 2001, promotes environmental literacy and develops a future science, technology, and engineering workforce, particularly from underrepresented communities. It is the largest program with a specific focus on broadening participation in the fields related to NOAA's mission (see Box 4.2 for descriptions of other initiatives that focus on broadening participation and interest).

The program has four components. First, the Graduate Sciences Program trains minority and women candidates in NOAA-related disciplines

and youth education programs, family field experiences, professional development for teachers and youth leaders, teaching resources, bilingual outreach, internships, and career mentoring, all geared for diverse multicultural communities. The program has been designed to encourage better public understanding of specific ocean-related issues within sanctuaries and their connections to inland human activities. It promotes community participation to address these issues and motivates culturally diverse students to pursue careers in marine sciences and resource protection.

Broadening participation also includes increasing the representation of girls and women in the fields of science, technology, engineering, and mathematics. The University of Southern California Sea Grant Program runs a summer science program for middle school girls, which aims to provide students with an introduction to oceanography through a series of hands-on experiences and scientific simulations. It demonstrates the effect that humans have on a delicate ecosystem through the study of marine biology and oceanography. The University of Southern California (USC) offers this program in conjunction with the USC Wrigley Institute for Environmental Studies and the Sea Grant Program. Sea Grant developed the program and scientists from the Wrigley Institute and the Sea Grant Program are the camp instructors and staff. This field-oriented program challenges students to explore ecological and biological principles through an interactive approach to learning, which includes field and laboratory research, self-guided experiments, and educational simulations. Science is integrated with cultural and social studies to offer a well-rounded view of an island ecosystem. A community of scientists is built through team activities, instructional and recreational activities, and mentoring programs. Students interact with faculty and graduates who are studying various aspects of science. Students also explore diverse careers in environmental, biological, and physical sciences.

and has paid for 21 graduate students' tuition, housing and other expenses, and 16 weeks of NOAA work experience per year. Second, EPP supports an undergraduate scholarship program that has provided NOAA internships and one-on-one mentoring to 66 students from underrepresented communities graduating with degrees in fields integral to NOAA's mission. Each scholar receives approximately $28,000 for two academic years to cover the costs of tuition and other expenses. Third, EPP has established five Cooperative Science Centers at minority-serving institutions. The aim is to build sustainable capacity via improved curricula and degree programs, increased laboratory facilities, enhanced national reputation, and a pipeline of students trained in science and engineering. The final component, the Environmental Entrepreneurship Program, provides competitive awards to minority-serving institutions to run initiatives geared to entrepreneurial or pipeline goals.

EPP Quick Facts
Scientific Focus—All NOAA critical disciplines
Audience—Undergraduate and graduate students at minority-serving institutions
Format—Scholarships, grants, and Cooperative Science Centers
Education Budget—$14.2M in FY2007
Line Office—Office of Education

Funds are used to either engage students and faculty in collaborative, field-based learning experiences to develop to business opportunities in their local communities, or to increase high school students' understanding of concepts related to NOAA's mission. The program has awarded $15 million to support 51 initiatives involving 1,460 students since its inception.

Impact: EPP support to the Cooperative Science Centers has resulted in 489 degrees awarded to students in core sciences related to NOAA's mission. In addition, EPP significantly increased the number of African American Ph.D. graduates in atmospheric and environmental sciences. In addition, the centers supported 62 new faculty members, over 150 collaborative research initiatives involving NOAA and minority-serving partners, and more than 200 peer-reviewed publications. The data received by the committee do not show number of degrees earned in various subdisciplines of ocean sciences or atmospheric sciences or in other specific subdisciplines related to NOAA's mission.

Overall, the evaluations found that the Cooperative Science Centers had made progress toward achieving their goals. EPP has demonstrated considerable success at increasing the number of marine and atmospheric science Ph.D.s received by underrepresented minorities. Moreover, it is very encouraging that many of these graduates (33) have gone on to careers as NOAA scientists. Evaluation of the entire EPP program is currently ongoing, yet each of the Cooperative Science Centers was evaluated in 2004 by NOAA and center staff. These evaluations took place in only the second year of the centers' operation and resulted in a number of recommendations to ensure future success. As described in Chapter 3, it is crucial that minority students have access to high-quality resources, facilities, and mentoring. Access to these resources is more likely when minority-serving institutions partner with majority-serving research universities and institutions. Continued tracking of the impact of the EPP program would be well served by establishing benchmarks and clearly articulated target diversity goals.

National Sea Grant College Program

The National Sea Grant College Program (NSGCP), established through the National Sea Grant College and Program Act of 1966, has matured into a state-federal partnership with a distinctive role and management structure. It is a nationwide network (housed in the Office of Oceanic and Atmospheric Research) of 30 individual Sea Grant programs. The Sea Grant infrastructure is modeled after the Land Grant system and includes partnerships with universities, a performance-based evaluation process, regional networks, and locally based infrastructure. The programs have over 300 state, federal, and industry partners. This network engages in conducting scientific research, education, training, and extension initiatives designed to increase assessment, development, utilization, and conservation of coastal resources by providing assistance to promote responsive research and training activities. In 2007, $53.3 million in Sea Grant awards were distributed: $23.9 million (45 percent) of the funds were for outreach and education, and $5.4 million was education specific.

Education is a key component of the NSGCP. It supports undergraduate and graduate education, teacher education, K-12 curriculum development, and informal education activities. Sea Grant has supported education and training of many marine and Great Lakes scientists, resource managers, and policy specialists through its three fellowship programs, the John A. Knauss Marine Policy Fellowship, the Sea Grant/NOAA Fisheries Graduate Fellowship, and the Sea Grant Industry Fellowship Program. In 2007, over 1,700 undergraduate and graduate students have received fellowships, scholarships, or research assistantships from Sea Grant.

In 2007, there have been 2,726 K-12 teachers and 1,516 informal educators involved in Sea Grant professional development programs through the National Sea Grant Educators Network. During that year the network funded 1,793 camps, programs, activities, and clubs involving 153,146 children and families to improve environmental literacy, and network educators taught 1,446 class field trips involving 139,259 students. These educators

NSGCP Quick Facts

Scientific Focus—Ocean and coastal science
Audience—K-12 students and educators, undergraduate and graduate students, and general public
Format—Varied, formal and informal education, fellowships, professional development, and curricular development
Education Budget—$5.4M in FY2007
Line Office—Office of Oceanic and Atmospheric Research

typically have academic preparation in both science and teaching. Most have advanced science degrees and formal or informal teaching experience. All Sea Grant educators are associated with universities or research institutions. In addition, curricular materials have been developed to promote ocean literacy. Typical Sea Grant curricular materials offer hands-on science lessons, often including new technologies that incorporate real-time data obtained from satellites and offshore monitoring instruments. Sea Grant offers many online resources to students and educators via the Bridge Ocean Sciences Education Teacher Resource Center.[4]

Impact: The Sea Grant review process provides periodic program assessments of each Sea Grant college, which include a review of education activities. Although many of the education activities are excellent, there is not much coordination among local programs. Better coordination across the Sea Grant network could lead to a greater impact. Evaluations of individual education activities show that most of them are received positively and that expected educational outcomes are being reached. For example, summative and formative evaluations of free-choice learning at the Hatfield Marine Science Visitor Center at Oregon State University led to small changes to exhibits that resulted in big changes in visitor comprehension, illustrating that age was not a barrier to the use of technology (Cammen, 2008). An evaluation of the USC's Summer Science Programs for Middle and High School Girls' Involvement showed that the program fosters interest in the environment and science, enthusiasm about what they learn, and higher expectations for classroom performance (Cammen, 2008).

During our Chesapeake Bay area site visit, we visited South Carroll High School in Sykesville, Maryland, where the Maryland Sea Grant (MDSG) has implemented Aquaculture in Action, a program to support aquaculture as a tool for teaching science. Through a partnership with the Carroll County school district, other local school districts, and local science organizations, MDSG provides teacher training, an online data system and teacher networking site, and materials to create an aquaculture laboratory. With support of MDSG and its partners South Carroll High School has created its own aquaculture laboratory (complete with recirculation tank system and computer lab). The program was created and implemented through the initiative of MDSG staff (with no education budget) and the support of its sponsors. The impact of these programs has not been evaluated, in part since there is no money for evaluation, but it is clear that the Aquaculture in Action program has created a sophisticated science classroom and provided teacher training that would not have been available if the program did not exist.

[4]See http://web.vims.edu/bridge/svr=www [accessed May 2010].

Staff creating, developing, and supporting such efforts with no or a very limited budget seems to be the rule rather than the exception in the programs we visited. There is a need to create a network or process for sharing effective practices in developing programs under these constraints, so that other programs and projects can benefit from the successes and failures of prior efforts.

Cooperative Institutes

NOAA funds 21 Cooperative Institutes (CIs) that conduct research on topics ranging from coastal ecology and fisheries biology to atmospheric chemistry and satellite climatology. The institutes, located at 34 research universities in 17 states, receive funding from NOAA through a competitive grants process to support both their research and their educational missions. Individual line offices manage the awards and oversee each CI's performance. For NOAA, the purpose of the partnership between research universities and NOAA laboratories is to promote research, education, training, and outreach aligned with NOAA's mission goals.

Although their primary work is conducting research, CIs also engage in a wide variety of educational activities. For example, the CI for Research in Environmental Sciences (CIRES) at the University of Colorado at Boulder has a graduate program in environmental sciences, an interdisciplinary outreach program, and online education resources for K-12 school districts, teachers, and students. K-12 education and outreach initiatives include classroom and prospective teacher professional development, volunteer opportunities for scientists, education components for research initiatives, district partnerships, and research mentors for high school students and undergraduates. Also, CIRES runs the Cooperative Program for Operational Meteorology, Education and Training, which promotes a better understanding of mesoscale meteorology and addresses education and training needs in the atmospheric and related sciences.

CI Quick Facts

Scientific Focus—All NOAA critical disciplines

Audience—Mainly undergraduate and graduate students, some K-12 and general public programs

Format—Competitive grants to universities

Education Budget—Not available

Line Offices—National Environmental Satellite, Data, & Information Services; National Marine Fisheries Service; Office of Oceanic and Atmospheric Research

Impact: In 2004, the NOAA Science Advisory Board conducted a review of agency research activities, which recommended, in part, that NOAA develop an agencywide policy for managing all CIs under a common procedural structure (Moore et al., 2004). We were not able to find out what education activities are supported across the CIs, as the tracking of education activities is inconsistent. Also, there has not been an evaluation of the overall education programs at the CIs, and the committee did not have access to evaluations of individual initiatives.

Ernest F. Hollings Scholarship Program

The Hollings Scholarship Program, run by the Office of Education, provides undergraduate students with awards that include academic assistance (up to a maximum of $8,000 per year) during the 9-month academic year; a 10-week, full-time internship position ($650/week) during the summer at a NOAA facility; and, if reappointed, academic assistance (up to a maximum of $8,000) for a second 9-month academic year. The purpose of the internship after the first year of the award is to provide scholars with hands-on educational training experience in NOAA-related science, research, technology, policy, management, and education activities. Awards also include travel funds to attend a mandatory Hollings Scholarship Program orientation, conferences at which students present a paper or poster, and a housing subsidy for scholars who do not reside at home during the summer internship.

Hollings Quick Facts
Scientific Focus—All NOAA critical disciplines
Audience—Undergraduates
Format—Scholarships
Education Budget—$3.9M in FY2007
Line Office—Office of Education

Impact: Since its inception, in 2005, scholarships have been awarded to 434 students. Several former Hollings scholars have been hired by NOAA.

Dr. Nancy Foster Scholarship Program

The Dr. Nancy Foster Scholarship Program, authorized in the 2000 reauthorization of the National Marine Sanctuaries Act, recognizes outstanding scholarship and encourages independent graduate-level research—

particularly by female and minority students—in oceanography, marine biology, and maritime archaeology. The program is administered through NOAA's Office of Education and funded annually with 1 percent of the amount appropriated each fiscal year to carry out the National Marine Sanctuaries Act. A maximum of $84,000 may be provided to master's students and up to $168,000 may be provided to doctoral students. Recipients are required to participate in a research collaboration at a NOAA facility. The research collaboration opportunity is designed to allow scholars to conduct their research at a NOAA facility and on NOAA mission research for four to six weeks. There were nine recipients of the Foster scholarship in 2008.

Foster Quick Facts
Scientific Focus—Oceanography, marine biology, and maritime archaeology
Audience—Graduate students
Format—Fellowships
Education Budget—$300,000 in FY2007
Line Office—Office of Education

Impact: A total of 30 students have received funding since the scholarship program's inception in 2000.

Additional Education Initiatives

Climate Communication and Education Program

Situated in the Office of Oceanic and Atmospheric Research, the Climate Program Office conducts the Climate Communication and Education Program (CCEP), whose mission is to improve public climate science literacy and to raise public awareness and understanding of and engagement with NOAA's climate science and services programs. CCEP was formed in 2008 and embarked on a multipronged new initiative to produce and distribute a range of products, conduct programs, and collaborate in partnerships designed to help NOAA fulfill its climate mission. It seeks to engage many segments of society, including policy makers, scientists, educators, and the public, to improve climate literacy. CCEP and its many partner agencies and organizations developed the Climate Literacy: The Essential Principles of Climate Science framework to guide the development of formal and informal climate science education curricula and supporting educational professional development.

CCEP Quick Facts
Scientific Focus—Climate science
Audience—K-12 students and educators and general public
Format—Varied, literacy principles, curricular materials, web resources
Education Budget—$4.6M in FY2007 for climate mission goal
Line Office—Office of Oceanic and Atmospheric Research

Impact: Most CCEP initiatives and products are still in the development phase, so it is too early to evaluate their effectiveness. The climate literacy principles provide an important framework for future curriculum development and formal education efforts.

Interagency Collaborations

NOAA is a sponsor of several interagency education initiatives. It contributes partial support to the National Science Foundation's (NSF's) Centers for Ocean Science Education Excellence (COSEE) network. This network was developed to promote a better understanding of the key role the ocean plays in global environmental cycles and processes. The network is comprised of 12 thematic and regional centers and a central coordinating office. The centers strive to foster the integration of ocean research into educational materials, enable ocean researchers to gain a better understanding of educational organizations and pedagogy, enhance educators' capacity to teach ocean science, and promote deeper understanding of the ocean and its influence on quality of life and national prosperity.

NOAA is one of eight sponsors and partners of the JASON Project. A nonprofit subsidiary of the National Geographic Society, the JASON Project uses the research of NOAA and the National Aeronautics and Space Administration to create science curricula on such topics as ecosystems and extreme weather. JASON attempts to use educational telepresence to create a "being there" experience for students to work side by side with scientists and researchers on real-world missions. A multiyear evaluation (Center for Children and Technology, 2003) found that JASON has had a positive impact on students' and teachers' understanding of science concepts and that it positively influenced students' perceptions of scientists and of becoming scientists.

NOAA is also a cochair of the Interagency Working Group on Ocean Education, established by the Interagency Committee on Ocean Science and Resource Management Integration. The working group is tasked to implement recommendations of the U.S. Ocean Action Plan to collaborate across federal agencies in order to increase ocean literacy and build a future workforce. Formally established in 2006, the working group has been meeting regularly to compare agency-funded programs and identify common priorities. The group is particularly focused on coordinating formal and informal education programs, developing a coordinated ocean message, promoting the use of ocean observation data in education, and attracting a future workforce to marine science, technology, and management.

NOAA has worked collaboratively with the National Geographic Society, the Centers for Ocean Science Education Excellence, and the National Marine Educators Association to create *Ocean Literacy Principles*, a document that defines ocean literacy. It is a resource developed to redress the lack of ocean-related content in state and national science education standards, instructional materials, and assessments. NOAA also co-funded a similar framework with NSF on atmospheric science literacy. In addition, NOAA collaborated with the U.S. Environmental Protection Agency and the North American Association of Environmental Education on the National Environmental Literacy Assessment Project to create a national measure of environmental literacy.

The Sant Ocean Hall, a collaborative effort of NOAA and the Smithsonian Institution, combines 674 marine specimens and models, high-definition video experiences, exhibits, and new technology, enabling visitors to explore the ocean's past, present, and future. NOAA invested $12 million in FY2006. The museum expected 7 million visitors in its first year. The hall contains 10 galleries, with exhibits on marine diversity, ocean exploration, ocean conservation, salmon and people, global ocean systems, and ocean evolution. There is also a Science on a Sphere exhibit and an Ocean Today kiosk, an interactive exhibit developed by NOAA specifically for the Sant Ocean Hall.

CROSS-CUTTING ISSUES

The program descriptions above illustrate that NOAA's education activities are as varied as the offices that implement them. The impact of the majority of programs is unknown due to the lack of reliable evaluation data. For programs for which reliable evaluation data of specific projects exist—such as the Educational Partnership Program, the National Estuarine Research Reserve System, B-WET, Environmental Literacy Grants, and the National Sea Grant Program—the results have been positive. Although it is impossible to simply summarize whether the combined efforts of the projects within any program have been successful or whether the combined efforts of the all programs are effectively reaching the agency's education goals, some general observations can be made about the education portfolio. Below we summarize our observations in discussing three cross-cutting issues: core ideas, web-based resources, and portfolio balance.

Core Ideas

The instructional activities in the NOAA education programs are generally focused on smaller scale concepts (i.e., the impact of harmful algae blooms) or environmental stewardship principles. Rarely did the instructional activities reviewed connect to big ideas in science or engineering or essential principles described in the oceanic, earth, climate, or atmospheric science literacy documents. For example, the Ocean Explorer instructional activity The Sea with No Shores is focused on the Sargasso Sea and has the following stated learning objectives:

- Students will be able to infer why the brown alga, Sargassum, is likely to be home to many marine organisms.
- Students can infer that the populations of organisms in the Sargassum are dependent on each other for survival (Ocean Explorer, 2002).

This activity for grades five and six could be aligned to larger scale science learning objectives, such as systems, diversity, or adaptation. Learning activities that have objectives that are clearly focused on big ideas in science provide the tools for student to apply these concepts across many other applications. For example, the systems concept could be applied by students to better understand economic, political, or social systems. The concepts of diversity and adaptation are the beginning of a scaffold for students to later understand concepts of evolution.

A number of the big ideas in science and engineering that are referred to in the *National Science Education Standards* (National Research Council, 1996) and the *Benchmarks for Science Literacy* (American Association for the Advancement of Science, 1993) are related to NOAA's mission. Concepts such as systems (e.g., the Earth as a system, atmospheric systems, oceanic systems, ecosystems), change (e.g., tides, weather, shoreline erosion), structure and function (the webbed feet of frogs living in estuaries, the long legs of wading birds, the bills of plovers), and matter and energy (e.g., carbon cycles and the ocean, El Niño) are just a few examples of big ideas in science. The Earth, oceanic, climate, and atmospheric science literacy documents also describe a set of important concepts in disciplines related to NOAA's mission. The literacy documents are useful tools; however, most teachers are not familiar with them, nor have they been widely adopted by school districts. The literacy documents define essential principles in specific areas of scientific literacy, rather than a set of core principles or ideas that are central across scientific disciplines. Thus, the literacy documents may be useful in informing decisions on what oceanic, earth, climate, or atmospheric science should be included in state science standards.

Connecting science education activities to core scientific concepts is key because that allows for instructional sequences that build students' understanding in a progressively complex fashion, enabling creative links to be made between disciplines (National Research Council, 1999, 2007). Thus, learning activities with instructional objectives that are clearly focused on big ideas in science or engineering can give students the ability to apply these concepts across many other applications, such as understanding desert, forest, or estuary ecosystems. Connecting to big ideas is a critical step in creating coherence across NOAA's educational materials, which cover a broad range of scientific areas. The agency's contributions to the oceanic, climate, and atmospheric science literacy documents are a promising first step in developing coherence.

Developing materials that align with core ideas is not easily accomplished. NOAA will need to seriously consider what steps can be taken to achieve this goal from a wide range of options, including creating new materials, adapting existing materials or providing professional development for teachers to better understand how to link existing materials to core ideas. Decisions on core ideas should be driven by the intersection of the agency's internal priorities and the educational needs of the nation. They should be guided by existing documents that describe core principles, such as the *National Science Education Standards*, the *Benchmarks for Science Literacy*, and the earth, oceanic, climate, and atmospheric literacy frameworks.

Web-Based Resources

An in-depth analysis of the agency's use of web-based technology was not possible given the vast array of project websites and electronic resources and paucity of information about their impact. It was apparent that many NOAA offices post educational materials on the web. For example, the Office of Education website includes links for students, teachers, and the public.[5] The links lead to websites with educational content on such topics as climate change, oceans and coasts, weather, and satellites and space. Almost all of the line offices include similar links to NOAA and other materials. In addition to outside links, the National Ocean Service has developed its own suite of online educational resources for students and teachers in partnership with the

[5]See http://www.education.noaa.gov/students.html [accessed May 2010].

National Science Teachers Association. Offerings focus on different aspects of oceans, corals, estuaries, and other marine topics and include tutorials, games, lesson plans, and teacher professional development.

Some programs have developed more sophisticated websites and online materials that allow students and teachers to engage with NOAA datasets and scientists in the field. Data on the impact of these sites were not available. However, the potential impact of these types of resources is great, because they have the potential to engage students and teachers in scientific explorations in meaningful and engaging ways. In addition, the development of well-designed web-based resources can expand the reach of NOAA education efforts. The potential impacts of these resources are more likely to be achieved if they are developed and implemented in connection with experts on these technologies and if research findings on effective use of these resources are implemented. It seems that NOAA has taken steps to connect with appropriate experts in some cases, but these connections do not seem to be consistent across programs.

The educational web presence of the agency can be improved. Website design is inconsistent across offices, many sites are low-tech, resources are difficult to locate, and some are redundant across offices. A well-designed suite of sites would allow for more effective browsing by activity and should include a robust search function (across NOAA offices) for easy navigation. The National Ocean Service website has the most materials and is also the easiest to navigate. Seeking guidance from website development experts inside and outside NOAA could greatly improve the accessibility and appeal of the agency's education websites.

Partnerships

NOAA education programs engage in a variety of partnerships. The two most common approaches observed were direct service partnerships and grants-based partnerships. Direct service partnerships typically occur when a unit of NOAA is responsible for organizing and presenting an education program. The term partners in this context refers to other education service providers (which might be other NOAA education programs, regional and local nonprofits, or education organizations) that may be called on to assist in the delivery of the program and the schools and organizations that send students (K-12 or adult) to participate in the education program. Grant-based partnerships were observed when a unit of NOAA solicits proposals from outside organizations that conform to the goals of the education program. The recipients of these grants and the participants in the resulting sponsored activity were also referred to as partners. Many of NOAA's education programs seem to pursue a mixed portfolio of direct service and grant-based partnerships.

NOAA has recognized the importance of partnerships as a means for achieving its strategic goals. This is an outgrowth of two factors: (1) the key role that the agency plays in the advancement and dissemination of NOAA-related sciences, and (2) the relatively small size of NOAA education programs compared with the needs for science education in the nation. Partnerships have been an important implementation strategy discussed in both the 2004 and the 2008 strategic plan.

However, NOAA has not yet developed criteria for guiding individual education programs in considering the types of partnerships that are likely to be effective for pursuing key strategic goals. Such decisions seem difficult to reach given the current organizational structure of its education programs, wherein each program operates relatively independently and has unique managerial structures, goals, and designs.

Nor have the programs developed guidance for the common challenges associated with implementing policy through partnerships. Partnerships often encounter a myriad of difficulties in practice that are often unanticipated by those who provide sponsorship, resources, and administration to the collaboration as well as by the collaborating agencies themselves (Bardach and Lesser, 1996; Hassett and Austin, 1997) especially when sponsors, administrators, and participants do not clearly understand the motivations and interests of participants (Hill and Lynn, 2003). Thus, as NOAA increases the level of coordination between its line offices and its internal and external partnerships, it needs to clearly define the goals and objectives of these partnerships.

When creating partnerships, NOAA should also follow practices for effective strategies for organizing partnerships, including identifying shared goals, designing experiences around issues of local relevance, supporting participants' patterns of participation (e.g., family structure, modes of discourse), and designing experiences that satisfy the values and norms and reflect the practices of all partners (National Research Council, 2009).

Portfolio Balance

As a federal agency with a mission that is not primarily related to education, NOAA cannot comprehensively address the nation's educational needs in areas related to its mission. Instead, the agency needs to balance how it makes use of assets to address national needs as well as its own. Thus, a balanced portfolio is not necessarily one that gives equal attention to all the critical factors listed below, but rather reflects intentional decisions on what critical factors to focus on. To make these decisions, NOAA will need to consider the needs of the audiences, its resources to address these needs, and the efforts of other agencies and organizations.

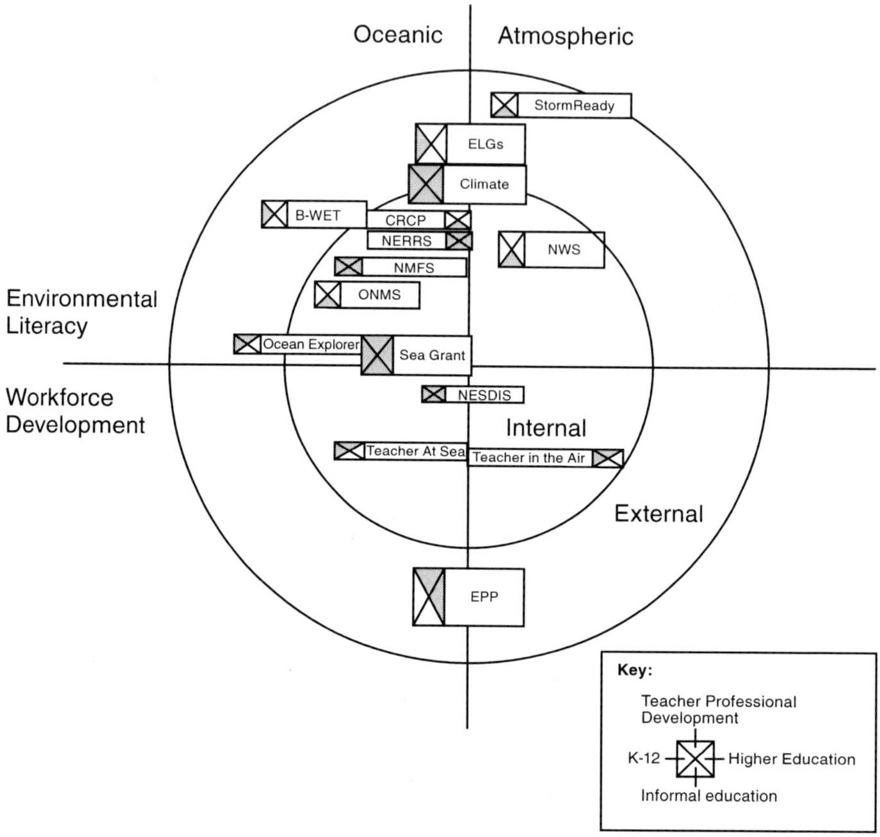

FIGURE 4.2 Portfolio balance.

The current portfolio balance is depicted in Figure 4.2. Three aspects of portfolio programs are displayed in two concentric circles. The inner circle represents programs that NOAA conducts internally. The relative amount of funding for each program is reflected in the size of its icon. The figure shows that the majority of NOAA programs focus on environmental literacy. Only a few programs focus on workforce development. Also, more programs focus on oceanic than on atmospheric science.

The number of internal programs is about equal to the number of external programs. However, programs vary greatly in budget and constituents served, so that the number of programs in a given area is not a reliable metric of balance, because it does not account for scope. For example, workforce development programs have large budgets compared with literacy programs. NOAA provides partial funding to many of the external programs it supports, and many of these programs also receive support from nonprofit organizations, foundations, and other federal agencies. NOAA is developing a system to collect more detailed information on its education portfolio that will make it possible to better understand its balance. However, the current portfolio does not seem to reflect any strategic portfolio-wide decisions regarding balance.

In thinking about portfolio balance, the context of supporting directives for education as well as the history of education at the agency must be considered. It should also be noted that many individual NOAA programs have their own educational mandates or are funded through congressional appropriations. These factors place constraints on how much NOAA can control the balance of programs in its portfolio. In developing a balanced education portfolio, NOAA must make decisions regarding the focus placed on six critical criteria, described below.

Balancing Environmental Literacy and Workforce Development

Although there are more programs that address environmental literacy, the budget for workforce development programs tends to be larger. The workforce goal is being tackled by a few large programs, whereas the literacy goal is being tackled by many smaller programs. This makes some sense, as the workforce goal is focused on a smaller, more focused audience of science, technology, engineering, and mathematics (STEM) undergraduates, graduates, and those in the pipeline to become undergraduates. However, without better information about the workforce needs of NOAA and the nation, it is impossible to judge whether NOAA is providing enough support for workforce programs, especially since it is likely now and in the future, as it has been in the past, that NOAA's workforce needs will span the broad range of STEM disciplines, not just oceanic, atmospheric, and climate sciences.

Balancing Oceanic, Atmospheric, and Climate Topics

As Figure 4.2 shows, there is currently a fairly large focus on ocean topics compared with atmospheric and climate topics. This is partly a consequence of the historical roots of NOAA education, with ocean-based programs, such as Sea Grant, NERRS, and the National Marine Sanctuaries, having clear mandates to engage in educational activities from their inception. The fact that NOAA's physical educational assets, like the national estuaries and marine sanctuaries, are ocean-based also contributes to this focus. However, there is no true dichotomy between oceanic science and atmospheric and climate science. There are clear connections between the ocean, the atmosphere, and climate, and education initiatives can be developed that integrate teaching about them. For example, ocean acidification can be used as an educational tool to make the connection between increases in CO_2 emissions in the atmosphere and the impact on the ocean. Such programs would more accurately reflect the true nature of the sciences related to these concepts, making it possible for NOAA to develop initiatives that address all three areas with its limited budget.

In addition, climate education is now a key priority for NOAA and the Obama administration, and funding for climate education in federal agencies is increasing. It is important that expertise be shared across the ocean, climate, and atmospheric disciplines so that new climate programming can benefit from lessons learned from the programs focused on ocean science, reflecting the integrated nature of these sciences.

Support for External Programs Versus Internal Programs

As mentioned earlier, NOAA can fulfill its role in education by using the educational assets it possesses and through collaborations with other agencies, organizations, and institutions. NOAA cannot be expected to house all of the necessary educational expertise to meet its educational objectives, just as it does not house all the nation's oceanic, atmospheric, and climate science and technology expertise. Even programs that are developed internally and led by NOAA can be aided by partnerships and collaborations with other agencies and organizations that have additional or complementary resources and assets. NOAA must be strategic in how and when to develop its education programs using intra-agency expertise versus supporting outside education programs whose developers may have more expertise in education and program design.

Encouraging Stewardship and Teaching Scientific Concepts

The environmental literacy goal in NOAA's strategic education plan contains a tension that is not explicitly addressed. NOAA defines environmental literacy as a fundamental understanding of the sciences and phenomena related to its mission and the use of that knowledge to make decisions about environmental issues. However, knowledge, behaviors, and decision making are not always linked. As mentioned earlier, behavior change and environmental decision making also encompass a number of psychological factors, including the belief that a person's decisions or actions can lead to change. Thus, behavior change will not flow solely from increased knowledge. To reach this goal, NOAA's education programs can use the effective practices in behavior change outlined in Chapter 3.

NOAA's focus on environmental literacy and stewardship also implies that education programs will have the goal of improving understanding of relevant STEM concepts, creating behavior change, or both. This presents some challenges. NOAA has not traditionally included behavioral change in evaluations of its programs and may not be well organized to analyze the behavior change of participants. And there is no shared, agencywide understanding of what it means to influence action. Influencing action can range from acknowledged and lauded attempts to influence behavior to enhance public safety (e.g., increasing hurricane preparedness) to advocacy on issues that might be regarded by some as controversial, like climate change.

Formal and Informal Learning Environments

Increasing environmental literacy and developing a STEM workforce both require a mix of programming targeted at the formal school system and informal experiences of young children, K-12 students, and the public. A recent NRC report (2009) highlighted the importance of informal education to inspire scientific interest and increase scientific competence in students, which can then translate into increased success in school settings. In addition, NOAA manages and protects place-based assets, such as the estuary reserves and the marine sanctuaries, which can be used for effective informal programming. Informal education does not mean that schools are ignored. Field trips and teacher professional development programs that occur in informal settings have clear connections to formal education institutions, educators, and students.

As discussed in Chapter 2, the formal education landscape is highly complex. NOAA should think strategically about how it can have a meaningful, positive impact in such a large system. NOAA can positively impact the education system by meaningfully aligning their instructional materials with the education standards of the states in which the programs are delivered. These materials should be supported with professional development that focuses teachers on important concepts from the national, state, and local environmental and science standards.

One interesting model for affecting the formal school system is in the Chesapeake Bay region, where the B-WET program supports informal education providers, yet requires all grantees to connect to a school system to ensure that the informal programs have a broad audience. This requirement is combined with policy work by the B-WET staff to win greater support for requiring outdoor education experiences for all school children among state and local governments. NOAA has a role to play in both the formal and informal sectors, and these sectors should be seen as complementary and intertwined.

Targeted Audiences Versus Self-Selected Participants

Many NOAA programs use self-selected participants. Data suggest that most programs are well received by these self-selected audiences. However, thought must be given to how to attract participation from traditionally underserved groups. To truly expand participation, interest, and understanding, the needs of multicultural audiences must be met across all education programs. Some efforts in this area, such as the Multicultural Education for Resource Issues Threatening Oceans program of the National Marine Sanctuaries, are promising, but too often broadening participation is not a key concern of the programs because of the self-selected nature of their audience. The vast majority of programs serve individuals who are already interested in the science related to NOAA's mission or schools that have the resources to support travel to NOAA education activities.

Breadth and Depth

An additional challenge for the education programs is to achieve a balance between projects that achieve a broad reach and those that foster deep engagement with the science and engineering content of the agency. The committee thinks that NOAA has an important role to play in both sorts of activities, which will require very different designs and deployment of resources. The agency's projects are therefore faced with striking a difficult balance between trying to make a broad impact while still providing meaningful engagement on a smaller scale. This balance can be partially mediated through modern technology, such as the Internet, which can be used as a distribution tool, and through strategic partnering with other federal science agencies and education organizations.

5

Current Evaluation Framework and Existing Evaluation Efforts

There is substantial evidence that the National Oceanic and Atmospheric Administration (NOAA) is giving increasing attention to evaluating the education and outreach programs that it supports. Evaluation is more prominent in the 2009-2029 education strategic plan than it was in the 2004 plan. Evaluation is primarily the responsibility of each individual education program, and the Education Council (EC) provides leadership and guidance for education evaluation activities across the agency. The EC has also developed and promoted an implementation framework incorporating program evaluation as an integral part of the management of program delivery. Program officers indicate that these efforts are in part a response to executive and legislative mandates as well as a perceived need to make better use of resources across individual programs and share lessons learned among the programs and their partners.

Evaluation can contribute to sound decisions on how to make strategic use of resources. Common metrics or program performance standards are evaluation tools that can support strategic decisions about resources. Evaluation can assist this type of decision-making process even in the absence of common metrics and performance standards. Programs will need to be compared in terms of two criteria: their strategic importance within the education portfolio and their relative effectiveness in achieving their stated outcomes. Lessons to share among programs and their partners can be derived from both formative and summative evaluation results (see Box 5.1 for definitions).

This chapter provides a review of evaluation efforts of the EC and individual education programs. Our review is based on presentations by NOAA

BOX 5.1
Formative and Summative Evaluation

Formative Evaluation: The purpose of formative evaluation is to provide feedback on the development of a program or project and its implementation. Formative evaluation results are used to make changes to programs during their development and initial implementation. An overarching formative question is "How is the project operating?" The specific questions focus on how the project is being implemented, including specific features of a project or program, such as recruitment strategies, participant attributes, materials, and attendance.

Summative Evaluation: Evaluation of a project's outcomes, also called summative evaluation, can be designed to address several questions. Summative evaluations are typically done after changes to the program have been made as a result of formative evaluations. They are used to determine whether, and to what extent, a program or project results in the desired outcomes.

program managers and key strategic partners as well as an archival review of 18 evaluation reports (10 internal evaluations, 8 external evaluations) of individual education programs (18 were selected from the 47 provided; these 18 included data on participants' reactions or outcomes and a description of the education program, participants, and data collection methods). In our review we observed considerable variance in the rigor of program evaluations. Most programs have not gone through a full outcome-based summative evaluation process; in fact, many have only recently begun to implement program evaluation at all. Even programs that have been in operation for many years and have significant performance measurement and evaluation procedures, such as Sea Grant, have difficulty in evaluating the scope and impact of their education activities. The norm among existing evaluations is not to measure impact, but instead to focus on such outputs as numbers served and the number of satisfied participants.

NOAA education managers demonstrated strong awareness of the limitations of past evaluation practices and have developed a plan of action in response. Therefore, rather than simply reviewing past evaluations, this analysis takes a forward-looking perspective, identifying the opportunities and challenges of linking existing formative and summative evaluation practice with the new strategies and goals set for NOAA's education initiatives.

We begin our review by examining the quality of individual program evaluations conducted in recent years. Our review of recent evaluations was assisted by two research papers prepared for this report. One developed a logic model of the education programs offered by NOAA (Clune, 2009).

The other is a review of 18 of the more substantive evaluation reports provided by NOAA education programs (Brackett, 2009). We follow this examination with a description of evaluation practices and reports that have achieved high regard within the agency.

Next, we describe the two most visible recent changes designed to improve evaluation practice. First, in 2007 the EC adopted the Bennett Targeting Outcomes of Programs (TOP) model as a framework for all evaluations (described below). In adopting the TOP model the EC is attempting to encourage individual programs to view evaluation as an integral part of program development, delivery, and management rather than an activity used to satisfy external accountability requirements. Second, in 2008 the EC led an agencywide initiative to develop a new strategic plan to guide the organization, coordination, implementation, and evaluation of the many education programs housed in NOAA.

CURRENT AND PAST PRACTICES IN PROGRAM EVALUATION

In this section we review the state of evaluation for education and outreach programs, based on evaluations that NOAA provided. The 2008 strategic plan and the adoption of the TOP model represent developments toward stimulating a stronger evaluation process in the education programs of NOAA. However, much can be learned from reviewing the variety of approaches that current education programs have employed in carrying out evaluations. We describe some general trends in evaluation among NOAA education programs, highlighting both the stronger and weaker practices.

The majority of evaluations tend to focus on local projects and initiatives, and programs have either not yet conducted a programwide evaluation or are in the process of doing so (Brackett, 2009). Table 5.1 summarizes the evidence indicating whether or not evaluations were available for this review. For the programs providing evaluations, the table indicates whether they were programwide evaluations (i.e., the evaluation attempted to collect evidence from actors across all sites or projects) or local assessments of individual projects or sites. The evaluations of six of the programs took a local perspective, examining individual projects or sites, and four took a program-wide perspective.

In evaluations that examined activities from a programwide perspective, the evaluation questions tended to focus on specific engagements with participants and their reactions to the experience. In our review of the evaluation reports, we examined the transparency of the reports with regard to the methodology and questions used in the evaluation. By transparency we mean whether the reader can see the questions asked of respondents in an evaluation, the methodologies chosen, and the characteristics of the

TABLE 5.1 Summary of Evidence on Evaluation Practices

Place-Based Programs Primarily Supporting Local Education Efforts	
Bay-Watershed Education and Training	Evaluation: programwide and local
Coral Reef Conservation Program	Evaluation: local
StormReady and TsunamiReady	No evaluation available
Place-Based Programs Primarily Conducting NOAA On-Site Education	
National Estuarine Research Reserve System	Evaluation: programwide and local
National Marine Fisheries Service	Evaluation: local
Office of National Marine Sanctuaries	Evaluation: local
Programs Focused Primarily on National Curriculum Development or Teacher Professional Development	
Environmental Literacy Grants	Evaluation: local
Science on a Sphere	Evaluation: local
JASON	Evaluation: programwide
Ocean Explorer	Evaluation: programwide
Teacher at Sea	No evaluation available
Programs Focused Primarily on Higher Education	
Cooperative institutes	No evaluation available
Education Partnerships Program	Evaluation: local
Ernest F. Hollings Scholarship Program	No evaluation available
Dr. Nancy Foster Scholarship Program	No evaluation available
Sea Grant	Evaluation: local
Other Initiatives and Interagency Collaborations	
Climate Communication and Education Program	No evaluation available
National Weather Service	No evaluation available
Ocean Hall/Ocean Kiosk	No evaluation available
Ocean Science Bowl	Evaluation: programwide

NOTE: An indication of "no evaluation available" should not be read as an absence of evaluation activity, but rather there was no evidence for making an assessment of the current practice.

participants. We also examined the degree to which measures aligned with outcome categories or concepts that serve the strategic needs of NOAA.

In addition to the items or questions in the evaluation metrics, Brackett (2009) notes the importance of transparency with regard to the larger hypotheses or guiding questions. These are "the key organizers of a strong evaluation, dictating the design of the study, the data collection strategies and instruments to be used, and the data analysis. The findings of the evaluation provide answers to these questions and the basis for interpretation of findings and recommendations" (p. 4).

TABLE 5.2 Focus of Evaluation Questions in 18 NOAA Program Evaluation Reports (number of reports)

Student learning/achievement (8) Student stewardship (6) Student interest in science, careers (3) Student satisfaction with activity (2) Student engagement in learning (1) Student sense of place (1) Student leadership (1)	Teacher learning (4) Teacher confidence in teaching ocean science (4) Teacher satisfaction with professional development (4) Teacher implementation or intent to implement practices, use materials (4) Teacher technology skills (2) Teacher stewardship (1) Teacher sense of place (1)
Scientist satisfaction with activities (2) Scientist learning (1) Museum visitor understanding/learning (3) Museum visitor satisfaction (3) Museum visitor suggestions for improvement (3)	Professional development provided, strategies used, program design (5) Professional development evaluation used (2) Program work environment (1)

SOURCE: Brackett (2009, p. 5).

By the standard of transparency, most of the evaluations reviewed fared well. Over three-quarters of the evaluations either provided the questions used or gave a strong enough implication of the nature of the question for the reader to understand what was being asked. A summary of the types of evaluation questions used across the 18 reports is presented in Table 5.2 and notes a suitable level of transparency. Although this suggests some room for improvement across programs, the norm for NOAA programs is to have acceptable levels of transparency in evaluation questions. In each quadrant of Table 5.2 the questions are aimed at different stakeholder communities important to the NOAA education mission. The 18 evaluations are aimed at understanding whether students, teachers, and museum visitors are learning and using the new knowledge they have been exposed to in their activities. There are also assessments of the participating scientists and their satisfaction in engaging with an education activity.

We also examined evaluation questions from the perspective of how well they served the strategic interests of NOAA. From this perspective, the evaluations also do good service. It is important to keep in mind that these evaluations were conducted under the guidance of the 2004 strategic plan, which placed greater emphasis on dissemination of NOAA science through the education and outreach programs. Brackett (2009, pp. 5-6) comes to a similar conclusion:

> A previous NOAA strategic plan emphasized the importance of getting NOAA science in use through the NOAA education programs. As evident above, none of the program evaluations reviewed indicated an evaluation question or program objective directly focused on use of NOAA science. It should be noted, however, that the use of NOAA scientific research and researchers was an underlying piece of most of the programs evaluated. Specifically, 15 of the 18 reports indicated in some way use of NOAA science and/or researchers as part of the program's work. Fourteen of the 18 reports provided teacher, student, or scientist satisfaction data concerning provision of the NOAA science research or data, or involvement of scientists in learning activities. Seven of the reports noted measures (self-report, tests, student presentations, or use of NOAA data) of teacher or student learning of NOAA-provided science content. Three of the reports gave no indications of a focus on using NOAA science, although one of these did provide a recommendation to develop a program using a system's research information.

This assessment offers some optimism that NOAA education programs will adapt to the new mission and develop evaluation questions to serve the needs of the strategic goals for environmental literacy and workforce development. The adoption of the TOP model, described in detail below, requires that evaluations now be guided by a different set of evaluation questions that align with the need to address the 2008 strategic education plan goals.

The design, data collection, and analysis of the evaluation reports reviewed for this study were of mixed quality. In briefings, NOAA program officers indicated that each program is ultimately responsible for monitoring the quality of the evaluations. We observed the result of this approach in the variance in evaluation design and quality. We also observed considerable variance in the reports themselves, with some elements being quite solid and other elements providing a weak foundation for reporting results and impacts. We highlight some of the stronger evaluation studies in the next section. Box 5.2 lists the strengths found in the evaluations, and Box 5.3 lists the issues of concern. NOAA program officers are well aware of the limitations of many of the previous efforts at evaluation and have taken steps to improve evaluation quality now and in the future.

Highly Regarded Evaluation Practices and Reports

NOAA program officers and NOAA documents highlighted examples of evaluation practices and reports that have achieved a high level of regard within the agency. These highly regarded evaluations have influenced NOAA's internal understanding of evaluation and shaped evaluation practices. These evaluations do not necessarily meet a high standard of evalu-

BOX 5.2
Notable Evaluation Strengths

- Reporting from multiple sources, such as teachers, students, staff, interns, scientists.
- Providing useful formative data and recommendations for programs and projects.
- Providing informative data on program impact.
- Effectively using and presenting descriptive statistical analyses.
- Effectively using and presenting inferential statistics.
- Rigorous, artful, and informative presentation of qualitative findings.
- Providing a particularly effective balance of quantitative and qualitative data.
- Including insightful literature reviews that were used to analyze program design and interpret findings concerning program implementation.
- Providing particularly clear, well-written overall reports.

SOURCE: Brackett (2009, p. 10).

ation quality, yet examining them in greater detail provided some insight into the views of NOAA officers about effective evaluation practices. We highlight the reports or processes because they represent a range of evaluation strategies and practices and illustrate the need for evaluations to be conducted and communicated in a manner that supports the uptake of evaluation findings.

Bay-Watershed Education and Training Program

In 2007, the Bay-Watershed Education and Training (B-WET) Program completed a large, external evaluation of the Chesapeake Bay area training program in Delaware, Maryland, and Virginia that included teacher and student data from many smaller projects in the area. This report is notable because of the efforts at providing a rigorous programwide assessment of performance and impact among students and teachers. The study provides extensive surveys of educators who partner with B-WET. The report also provides matched comparisons of student performance in classes who have and have not participated in B-WET programs.

The 2007 evaluation is also notable because it is the most rigorous evaluation design employed among the NOAA evaluation programs. During the development of this report pressure was growing in the federal government to incorporate more rigorous designs into all program evaluations. The B-WET evaluation study coincided with the work of the Academic Competitiveness Council and the release of its report advocating rigorous

BOX 5.3
Notable Evaluation Weaknesses

Clarity and Focus
- Lack of description of the program being evaluated.
- Lack of evaluation questions to focus the report.
- Lack of conceptual framework needed in some cases.

Methodology and Instrumentation
- Missing detail on methods, such as sample selection, questionnaire piloting, and administration.
- Some overdependence on self-report.
- Overdependence on either qualitative or quantitative data.
- Some questionnaires and interviews poorly constructed.
- Some questionnaire items designed more for data analysis than for comprehension by respondents or for finding out what they actually think (these were probably never pilot tested).
- Comparison study using a control group of students has very little information on the implementation of the program, except for the number of hours spent—which varies a great deal.

Data Analysis and Presentation of Results
- Some data collected but unreported (interviews, site visits) or unanalyzed (analysis across classroom observations).
- Poor or no analysis of qualitative data, such as giving a simple list of comments.

evaluation designs (U.S. Department of Education, 2007). At that time, "rigorous" was narrowly defined as clinical trials or equivalently designed evaluations, rather than the most appropriate for the evaluation purpose and questions. In public documents NOAA has touted the B-WET evaluation as an example of the responsiveness of the agency to the national policy initiatives emphasizing rigorous evaluation.

Sea Grant Program

Sea Grant, the oldest initiative in the EC, submitted 40 reports for our review of evaluation practices, but most were either assessments of a state's Sea Grant Program (with a description of what education activities occur and who is served) or results from post-program surveys (i.e., the percentage of people who provided each type of response to questions about the activity). Of these reports, three were evaluations of an education

- Poor presentation of results of basic descriptive statistical analyses, such as unclear graphs, missing sample sizes.
- Poor use of data from multiple sources, so that, although many different groups were surveyed, their roles and specific concerns were not made clear.
- Complex statistical analysis that may be more than what is needed for the purposes of the evaluation and answering the evaluation questions.
- Some statistical analysis apparently done mostly "for the sake of doing statistics," since these findings were ignored and not used to inform recommendations.
- Extensive statistical analysis using national data across programs yet presented pretty much in isolation, with no evaluation questions, no discussion, and no recommendations.

Interpretation of Results and Recommendations

- No reflection, interpretation, or discussion of findings in ways that might help programs improve.
- Use of unanalyzed or unreported qualitative data to make recommendations.
- Evaluator so focused on the conceptual framework of the program that he or she neglects to bring forth what was actually asked of respondents and what they had to say about the program, leading to an artificial analysis of data and weak recommendations.
- No recommendations presented at all.

SOURCE: Brackett (2009, pp.10-11).

activity that included appropriate and sufficient information to warrant review. One consisted of an internal online questionnaire of 46 members of the Sea Grant Network conducted in 2008 to gather information about the programs and their needs. The other two reports provided evaluation information on two separate teacher learning projects: Teacher Education at Stone Laboratory (Ohio, undated) and the Aquatic Invaders in Maine (AIM) Teacher Workshop.

What is most notable about the Sea Grant Program is the extensive formal performance monitoring process that it has developed. The performance monitoring process is a tool used with all Sea Grant projects and activities, including those aimed at educators. Thus, the evaluations of Sea Grant include all research and extension-related activities, which are beyond the purview of this report. Sea Grant leadership indicates that the most common use of evaluations is as a performance management tool that provides Sea Grant program officers with up-to-date information on the state of project implementation. Participants in Sea Grant programs are required as a condition of sponsorship to submit annual reports detailing progress to date.

Sea Grant uses this information every four years as part of a review of individual Sea Grant programs. This process for assessment has become a part of Sea Grant's standard operating procedures. It is the only process observed that provides a scheduled, project-oriented view of performance. The key criteria for assessing Sea Grant programs consist of ratings for organizing and managing the program, connecting Sea Grant with users, effective and long-range planning, and producing significant results.

Sea Grant leadership reports the following challenges associated with the current evaluation process after two rounds of program assessments. First, program assessments are broadly focused on the entire research program. Consequently, there is relatively little time or talent dedicated to examining the education and outreach programs. Second, program assessments tend to focus on one university program at a time. This means that there are few opportunities for a comparative assessment across programs. However, a strength of the Sea Grant approach to evaluation is that it may create an information infrastructure that can be used to integrate planning, implementation, and evaluation. There is not sufficient evidence to judge whether the information infrastructure has been used in this manner.

Office of National Marine Sanctuaries

The Office of National Marine Sanctuaries (ONMS) is interesting because of the leadership and advocacy that it has been providing in recent years toward improving the quality of evaluations across NOAA education programs. It is through ONMS that the EC was introduced to the TOP model and ultimately adopted this approach across all NOAA education programs.

At the time of our study, there was no ONMS evaluation report available that had fully incorporated the TOP model. ONMS is currently developing a programwide evaluation that employs this approach. What was available was a series of project evaluations detailing the implementation and impacts of specific engagements with students and teachers. Each of these reports evaluated the use of the marine sanctuaries as a living classroom. For example, a 2004 assessment of the Dive into Education program examined the professional development of 62 K-12 teachers in Hawaii and American Samoa as they developed national science education standards-based ocean science activities aimed at stimulating student learning. Evaluations were also conducted to assess participant satisfaction with the LiMPETS (Long-term Monitoring Program and Experiential Training for Students) program, which provides teachers with training in marine science protocols that can be applied in the classroom and the field. LiMPETS also provides students with an ongoing scientific process to monitor natural resources and contribute to databases over time. Other evaluated projects, such as the 2005 Hawaii Field Student, focus even more strongly on conservation and stewardship by following pairs of teachers and students as they interact with coral reefs and larger ocean ecosystems. We highlight these evaluations because the programs are viewed by NOAA as a good model for a program that achieves alignment with standards for teaching. As ONMS continues its implementation of the TOP model, it can test these assumptions.

Overall, it is clear that NOAA is engaged in various types of evaluation, that some programs have conducted evaluations of varying quality, and that the evaluations of higher quality could serve as models for other programs. For example, practices that are worthy of replication include recruiting a comparison group when useful and appropriate; creating an information infrastructure to integrate planning, implementation, and evaluation; collecting information from multiple sources; aligning evaluation questions with program goals; collecting both quantitative and qualitative data; and using literature reviews in early stages to understand best practices in program design and interpret program implementation findings.

Until NOAA articulates measurable goals and outcomes for its education programs, it will be difficult to design evaluation questions that align goals and outcomes or produce any summative results at the highest level. Instead, the agency is left primarily with formative results and, in a few cases, localized results showing impacts that serve as tests of the program design. Once a set of overarching measureable goals and outcomes is articulated, it will be possible to assess projects against those goals and outcomes. Such assessments are likely to reveal successes and failures, from which would emerge common metrics, instruments, and practices that could be promoted for use across similar types of programs (e.g., teacher training), providing the needed data for summative evaluation across NOAA programs. Decisions about the goals, outcomes, and assessment metrics should be made by NOAA staff with appropriate experts (program designers, evaluators, education staff from other agencies and institutions, among others).

THE ROLE OF EVALUATION IN THE 2008 STRATEGIC PLAN

While the 2004 strategic plan focused on translating NOAA science into useful knowledge for the education communities, the 2008 strategic plan is a more ambitious articulation of the agency's goal of addressing the environmental literacy and workforce needs of the nation, in line with authority given to it by the America COMPETES Act. Under each goal the importance of evaluation is stressed with respect to specific strategies for achieving key outcomes. In contrast, the 2004 strategic plan provided no specific mention of evaluation in the goals or strategy statements.

The 2008 strategic goals pose several challenges to existing evaluation practices. Perhaps the most significant challenge is establishing evaluation processes through which NOAA can assess the cumulative impact of education programs toward achieving the strategic national goals. The programs are numerous and relatively small in light of the mandated mission. The 2008 strategic plan articulates several key factors driving variability in the missions of the education programs and, subsequently, in the evaluation strategies pursued. Among these are authorizing legislation for the individual education programs, the diverse body of disciplines related to science, technology, engineering, and mathematics (STEM) on which the programs draw, and the target communities with which the program interacts (e.g., K-12 education, informal education institutions, postsecondary education).

A mismatch is noted between the 2008 strategic goals and the scope and scale of the evaluations conducted to date. Evaluations in NOAA tend to collect self-reported impact data and information about participant satisfaction. Larger evaluation questions about the effective allocation of resources tend not to be addressed, nor do the individual programs report an incentive for this type of assessment. The Office of Education managers explained that one way of addressing these problems is through the EC, which serves as a forum for sharing information across the portfolio of education programs. Current efforts to develop an implementation plan to support the 2008 strategic education plan include formalizing comparative reviews of evaluations as a part of the work of the EC.

While sharing program evaluations is a positive step, it does not address the issue of conducting appropriate evaluations that allow the EC to determine which of the education programs are effective and what parts of these programs contribute to success. It is also difficult to see how sharing the results of evaluations will provide a sufficient foundation of information to guide in the strategic allocation of education resources. In order to accomplish this, NOAA would have to weigh questions of value (whether a certain type of essential outcome is being addressed, such as document analysis) with questions of effectiveness and efficiency (such as outcomes-based evaluations that include an appreciation of input variables). Highly effective programs may not address particularly important strategic goals; conversely, programs in need of substantial improvement might be uniquely positioned to address them.

To further illustrate our concern regarding the mismatch of the scope and scale of evaluations, we turn to the paper prepared for the committee by Clune (2009). Its logic model for NOAA education programs (see Table 5.3) notes that the relative weakness of the central governing authority allows the following to occur:

- Redundancy of effort in the development and implementation of programs.
- Overlapping constituencies for programs.
- Barriers to promoting common standards for curriculum materials and pedagogy.
- Barriers to having common cost-benefit standards for determining the effectiveness of programs.

TABLE 5.3 Common Logic Model for NOAA Instructional Programs

Logic Model Elements	Corresponding NOAA components
Inputs	
	Educational goals in a research agency *provide guidance for:* Educational management *that creates and administers:*
Activities	
	Instructional activities *directed at:* An audience (or audience clusters) *consisting of:* Educational content, instructional materials, pedagogy *delivered at/through:* A geographical site, website, partnership *aimed at producing:*
Outcomes	
	Learning outcomes *knowledge about:* (1) Natural resources—Reefs, estuaries, fisheries, etc. (2) Negative human behaviors—Pollution, overuse, climate change, etc. (3) Stewardship—Ameliorative decisions, policies, conservation
Medium- and long-term outcomes and impacts	
	Behavioral outcomes, including positive: Decisions, policies, operations, politics *that lead to:*
Impacts	
	Societal outcomes, including: Conservation, restoration, sustainable use, and development

From an evaluation perspective, this logic model raises two questions: (1) Does NOAA have an adequate forum for addressing issues of redundancy, overlapping constituencies, etc.? and (2) Is a sufficient information base being collected to address these types of problems? The evaluations reviewed by the committee do not include these types of issues as a focus of inquiry. This is probably because existing evaluations were aimed at assessing individual projects or events in and of themselves rather than in comparison to one another or the overall program. Some capacity for these types of evaluations is needed for NOAA to be able to convincingly demonstrate that the collection of information on individual programs is bringing about the outcomes related to environmental literacy and workforce development. On that note, NOAA needs to provide intermediary goals that are more in line with available resources. The committee appreciates, for example, that NOAA cannot increase the environmental literacy of the entire U.S. population. That said, however, what exactly would be a realistic goal for which NOAA should be accountable?

A second evaluation challenge growing out of the 2008 strategic plan is the importance placed on partnerships. The plan identifies partnerships across NOAA programs, with other federal and state agencies and with formal and informal education institutions. The 2008 strategic plan further identifies 29 distinct strategies for achieving outcomes aimed at fulfilling the two strategic goals. Partnership and interorganizational collaboration are key components in 19 of these 29 strategies.

Partnerships have been a focal point for evaluations in school-university partnerships (Goodlad and Sirotnik, 1988) and STEM education programs (Scherer, 2008) as well as other policy domains (Brinkerhoff, 2002). However, there is limited evidence that current evaluations conducted by NOAA account for the influence of partnerships in achieving outcomes and impacts. In addition, most evaluations do not attempt to observe the underlying partnership or explore this as a factor in assessment. Given the strategic importance placed on partnerships, greater attention to this topic is needed across NOAA program evaluations.

A related evaluation issue is that the education strategic plan does not specify a role for scientists and engineers and science offices in education, so it will be difficult to formulate, justify, or enforce evaluation metrics to determine if the interplay between agency scientists and engineers and education staff is working. It is critical that NOAA scientist and engineers have a role in the education efforts, because, as discussed in Chapter 3, they are one of the important assets that NOAA can use to address its educational goals and the needs of the nation. Collaborations between education staff and scientists and engineers have the potential to lead to higher quality education programs and resources than would be possible without such collaborations. Thus, just as the contributions of NOAA education staff need to be evaluated, the contributions of its scientists and engineers need to be assessed so that their impact is not overlooked, so their contributions are appreciated at an institutional level, and so that continual improvement of their connection with education staff is possible.

A third evaluation challenge arising out of the 2008 strategic plan stems from the emphasis placed on the development of consistent performance metrics across programs aimed at improving environmental literacy. Performance metrics are to be applied in formal and informal education programs and for the many disciplines that contribute to environmental literacy. The EC is the suggested forum for sharing knowledge about the development of performance metrics and disseminating effective practices.

As in the challenge posed by partnerships, there is a significant gap between current evaluation practice and the goal of using common performance metrics. NOAA is making important investments in conducting the baseline research for constructing such metrics, as evidenced in the sponsorship of the National Assessment of Environmental Literacy conducted by the North American Association for Environmental Education (see McBeth et al., 2008). However, there is little evidence to date of the use of common metrics across NOAA program evaluations, nor has NOAA conducted a critical analysis to determine the feasibility of common metrics across significantly differing programs. There are enormous and probably prohibitive challenges in designing and applying common metrics in ways that would lead to comparable data and information. Even if theoretically possible and technically feasible, questions remain: Would common metrics be useful and desirable? Or could they lead to a centralization and homogenization of NOAA education programs, which could ultimately threaten the value of place-based, individual, local education efforts?

A fourth challenge is the emphasis of various education initiatives on reaching diverse communities. This emphasis must extend to the evaluation of NOAA's programs by using evaluations that are sensitive to the existing context of culture and diversity. Such evaluations consider cultural diversity at all stages of the evaluation process: selection of stakeholders, development of evaluation questions, design of the evaluation, data collection and analysis, and communication. Evaluations should be culturally responsive and operate in a manner appropriate for the audiences being served. Some aspects of a culturally responsive evaluation are showing genuine respect for participants and engaging in an ongoing process of awareness of contextualized cultural needs (Mertens and Hopson, 2006). This also includes thoughtful consideration of culturally enforced differential access and resource opportunities. This perspective may substantially affect the timing and conduct of an evaluation and help to uncover basic but unstated assumptions about programming or evaluation findings. Culturally sensitive and appropriate evaluations also may run counter to the ideal of common metrics and comparable evaluation results. The goal of developing culturally appropriate and sensitive evaluations may therefore be in conflict with goals for comparable evaluations based on common metrics.

A fifth challenge with conducting more and more rigorous and comprehensive evaluations is that evaluation, particularly when focusing on impact at the program level, can be expensive. Available funding for education programs and evaluations is limited. A rule of thumb for evaluating programs is that at least 5 percent of the total budget should be devoted to summative evaluation. Formative evaluation should be part of program design, and its cost is part of the program. Reports from project managers indicate that this level of funding for evaluation has not been provided. Insufficient funds severely limit the scope and nature of any evaluation. Given limited overall funds, it is critical that NOAA develop a plan for allocating the funds for evaluation.

To achieve the greatest return on limited resources, evaluation of individual projects can be scheduled on a cyclical basis, with high priority given to projects intended to have the greatest impact on environmental literacy and workforce needs and to projects that face important questions about activities, participants, staffing, funding, or organization. Both formative and outcome evaluations can usually be scheduled in advance. For example, reports about program effectiveness may be scheduled on a periodic basis: staff can plan for outcome evaluations in advance over a 4-5 year period, rotating the projects in the portfolio.

THE TOP MODEL AND EVALUATION

In April 2007 the EC adopted the Bennett TOP model (Bennett and Rockwell, 1995), a specific version of a logic or program model, as a common framework for evaluation (see Figure 5.1).

TOP focuses on outcomes in planning, implementing, and evaluating programs. TOP documentation presents itself as based on a hierarchy, integrating program evaluation in the program development process using a simple framework to target specific outcomes in program development to existing resources and outputs, and then to assessing the degree to which the outcome targets are reached. As with most evaluation frameworks, the TOP model does not provide guidance on specific methods or metrics for implementing individual evaluations.

TOP is described as based on a theoretically sound framework that has been tested, revised and refined, and widely used over the past 20 years (Bennett, 1975, 1979; Bennett and Rockwell, 1995). The model consists of seven progressive levels of outcome assessment derived through a system-

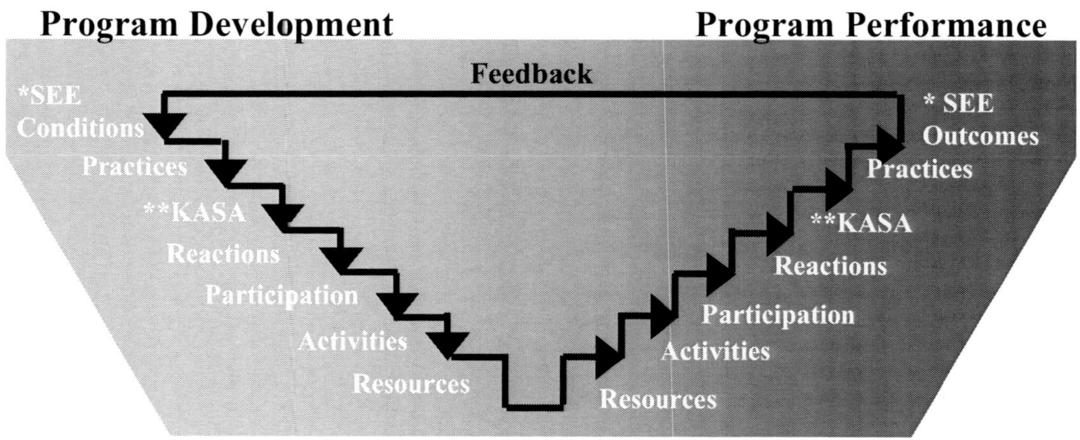

FIGURE 5.1 Bennett TOP model.
NOTE: *SEE, S = social, E = economic, E = environmental; **KASA, K = knowledge, A = attitudes, S = skills, A = aspirations.
SOURCE: Rockwell and Bennett (2004).

atic process of program development process. The seven assessment levels, presented on the TOP website, are briefly defined as follows:

- The **Resources** (1) level explains the scope of the programming effort in terms of dollars expended, intellectual resources, partnerships, and other assets.
- Progress documented at the **Activities** (2) and **Participation** (3) levels generally is referred to as outputs. It indicates the volume of work accomplished and is evidence of program implementation.
- The **Reactions** (4) level, is evidence of participants' immediate satisfaction.
- Intermediate outcomes at the **KASA** (knowledge, attitude, skills and aspirations) (5) level focus on knowledge gained/retained, attitudes changed, skills acquired, and aspirations changed.
- Intermediate outcomes at the **Practices/Behavioral** (6) level focus on the extent to which practices and behaviors of program participants are influenced. These outcomes can be measured months or years after program implementation, and they can also be accomplished in the short term, and may be measurable immediately following an intervention.
- Intermediate outcomes lead to longer term social, economic, and environmental (SEE) changes, or impacts of the program or activity. Identifying outcomes at the **SEE** (7) level (akin to defining impacts) for localities may occur fairly quickly although state, regional, or national outcomes may take years to assess and may be very expensive.

The developers say that the strengths of the TOP model are its focus on the educational process and incorporation of a broad range of outcome-based evaluation techniques, as well as the fact that the outcomes are aligned in accordance with theories of behavioral change. At the management level, the educational approaches of individual projects can be compared for their effectiveness in achieving similar outcomes.

Models for evaluation have been developed for many years, and many began in the 1960s with the growth of federal accountability. The purpose of models for evaluation is to provide a mechanism for covering the range of issues involved in programs and mechanisms for determining effectiveness. Each model emphasizes different aspects of educational programming. Along with the development of models has been the development of standards for evaluation. Current evaluation standards from the Joint Committee on Standards for Educational Evaluation (1994)[1] include

[1]An updated version of the Standards for Educational Evaluation Program will be released in 2010.

- **Utility Standards:** These standards relate to guaranteeing that the evaluation information will be used once it is completed. In order to accomplish this, suggestions about how to engage in the following activities are provided: stakeholder identification, evaluator credibility, information scope and selection, values identification, report clarity, report timeliness and dissemination, and evaluation impact.
- **Feasibility Standards:** These standards relate to guaranteeing that the evaluation can actually be carried out. Suggestions for how to conduct the following activities related to feasibility are provided: practical procedures, political viability, and cost effectiveness.
- **Propriety Standards:** These standards relate to guaranteeing that the evaluation is conducted in a fair and equitable manner. To ensure propriety, suggestions on how to conduct the following activities are provided: service orientation, formal agreements, rights of human subjects, human interactions, complete and fair assessment, disclosure of findings, conflict of interest, and fiscal responsibility.
- **Accuracy Standards:** These standards relate to guaranteeing that the evaluation information is valid, reliable, and analyzed appropriately. To ensure accuracy, suggestions on how to conduct the following activities are provided: program documentation, context analysis, described purposes and procedures, defensible information sources, valid information, reliable information, systematic information, analysis of qualitative information, analysis of quantitative information, justified conclusions, impartial reporting, and meta-evaluation.

The standards are based on the definition of evaluation as the assessment of something's merit. The standards can be applied to any evaluation plan or evaluation model to determine its quality. In addition to the standards, Daniel Stufflebeam, the first chair of the Joint Committee on Standards for Educational Evaluation, recently provided an assessment of evaluation models using the 30 standards (Stufflebeam and Shinkfield, 2007). Stufflebeam and Shinkfield suggest five categories of program evaluation models or approaches: (1) pseudo-evaluations, (2) question- or methods-oriented, (3) improvement or accountability, (4) social agenda and advocacy, and (5) eclectic.

The TOP model appears to fall into the improvement/accountability category, along with approaches such as the CIPP (Context, Input, Process, and Product) model (Stufflebeam, 2005) or that of Cronbach (1982). The central thrust of this type of evaluation is to foster improvement and accountability by informing and assessing program decisions. In considering the TOP model in terms of each of the standards, it appears to be a reasonable model for evaluation but limited in detail about how to actually implement it, although this detail might be available from a TOP expert or experienced evaluator.

The TOP model connection between evaluation and programming is especially valuable for program development. The far-reaching aspects of social, economic, and environmental changes also fit well with the stewardship goals of NOAA. The needs and opportunity assessments described in the TOP model appear to be quite similar to the context and input portions of the CIPP model. There are also several real-world examples provided on the TOP website to help operationalize the model.

However, the model may not translate well to broader cross-program issues or to programs that are less oriented to participant development. The TOP model does discuss interorganizational issues, offering a five-level approach of networking, cooperation, coordination, coalition, and collaboration. The more outcome-based orientation of TOP might not provide sufficient feedback for program improvement, although the emphasis on program development might compensate for the lack of feedback. In addition, there appears to be less emphasis on the utility standards than could be warranted. The TOP model is strong in terms of the feasibility standards. It is difficult to understand from the material provided how the propriety standards would be met. The accuracy standards are critical to any evaluation, and more emphasis on these standards would improve the TOP model.

Brackett (2009) examined 18 NOAA evaluation reports submitted for analysis for conformity to the TOP model using the seven levels of assessment. Brackett notes that many of the evaluation reports submitted were conducted either prior to or coinciding with the adoption of TOP model by the EC and the issuance of any guidance to the individual programs.

Although only a small proportion of education initiatives have been evaluated, "17 of the 18 [evaluation] reports provided some data concerning Program Activities and Participation, although this information was often spotty in nature. All 18 reports provided information on the Reaction level. Sixteen reports provided data concerning intermediate outcomes in the area of KASA. Nine included information concerning intermediate outcomes in the area of Practices or Behaviors. None of the reports provided information concerning broader SEE changes" (Brackett, 2009, p. 10).

Brackett's review suggests that the EC has good reasons for promoting TOP as a standard for evaluation by NOAA's education and outreach programs. Current evaluations tend to focus on the specific forms of engagement by participants in NOAA-sponsored programs as well as evidence of reactions or learning taking place. Application of the TOP model may offer a useful reminder to program officers that they need to stretch the scope of current evaluation practice to include the resource inputs and larger changes in practices and social impacts at the overall program level (not at the individual activity level). At the resource level, "staff time used" must include scientist time as well as educator time; it is not appropriate to expect scientists to squeeze education activities into their spare time while expecting them to carry a full load of scientific responsibilities.

Although the TOP model is not a magic wand for solving the evaluation challenges faced by NOAA, it does serve as important guidance reminding managers that evaluation is not simply an accountability chore on a checklist. With the proper scope, evaluations can provide critical information for program management. However, until there are real objectives in the strategic plan and good strategies for collecting data related to them, summative evaluation across the agency's diverse and loosely coordinated education portfolio will remain conceptually challenging. This is true for assessing progress toward both the environmental literacy goal and the workforce goal. For example, assessing progress toward the workforce goal would necessitate either long-term longitudinal data or a plan for what to do in the absence of such data. It might also be necessary to develop pipeline metrics that could illustrate whether programs address critical bottlenecks or leaks in the pipeline, especially for individuals from underrepresented groups.

SUMMARY

Evaluation of federally funded education programs is evolving rapidly, and at the same time the expectations of NOAA programs have changed quickly. The agency has responded and in some cases has done exemplary work.

NOAA is increasing its emphasis on evaluation. It is using the Office of Education and the Education Council to coordinate evaluation activity and is adopting the TOP model. The model is a reasonable one.

Although NOAA is conducting evaluations of its educational activities, they are limited in scope and tend to focus on immediate and intermediate outcomes. Nearly all evaluations lacked comparative elements. Most serious is that there is little consideration of evaluation at the portfolio level, such as between different programs or different approaches or in terms of what types of programming might be most effective in meeting NOAA educational goals.

NOAA can improve its evaluation strategy by:

- Increasing the emphasis on high-quality evaluations by using the higher order evaluation suggestions of the TOP model as well as its program development and improvement aspects.

- Incorporating effective practices into any evaluations that are implemented.
- Emphasizing evaluation of the entire portfolio of NOAA activities using consistent data-gathering approaches.
- Evaluating both the education programs and the line offices on how effectively the ideas, insights, knowledge, understanding, and passion of the agency's scientists and engineers, as well as other scientists and engineers in the relevant disciplines, are incorporated into educational materials and programs.
- Evaluating the appropriateness and effects of their partnerships.

6

Conclusions and Recommendations

This chapter presents the committee's conclusions and recommendations on five aspects of education programs at the National Oceanic and Atmospheric Administration (NOAA): (1) the agency's role in education, (2) its education goals and outcomes, (3) the composition and management of its education portfolio, (4) its education evaluation practices, and (5) the impact of its education efforts. The committee's conclusions and recommendations are based on the materials and testimony outlined in Chapter 1, the data and analysis presented in Chapters 2-5, and informed by the scientific, engineering, educational, and evaluation expertise of its members.

I. NOAA'S ROLE IN EDUCATION

The national need to educate the public about the ocean, coastal resources, atmosphere, and climate and to support workforce development in related fields is well established. The federal government has an important role in addressing these needs as part of the national effort to improve science, technology, engineering, and mathematics (STEM) and environmental education. Although NOAA's focus is unique, many other federal agencies also address the country's education and workforce needs related to the ocean, coastal resources, atmosphere, and climate. It is challenging to coordinate these activities in a cohesive way, making the best use of the assets of each agency as well as the infrastructure and capabilities outside the federal government.

NOAA's specific role in education is influenced by multiple factors. As a federal agency, it is responsible for the conservation and management of coastal and marine resources to meet the nation's economic, social, and environmental needs. It also promotes understanding and prediction of changes in the Earth's environment. NOAA's role in education is tightly connected to its mission, and some education programs are defined by various congressional mandates. NOAA must fulfill these responsibilities in the context of a national effort to improve learning and understanding of STEM implemented at state and local levels in both formal and informal settings for learning, and at the same time carry out effective environmental education.

Conclusion I.1: *The America COMPETES Act assigns NOAA responsibility for advancing and coordinating mission-related STEM education and stewardship efforts and for participating in interagency education efforts (Chapter 1: Overview and Introduction).*

NOAA's role in education has been recognized for more than 30 years, as evidenced by the legislative mandates to engage in education activities given to its individual operating branches and programs. This role was underscored and strengthened by the America COMPETES Act, which gives the agency a broad mandate to engage in and coordinate education initiatives in oceanic, Great Lakes, climate, atmospheric, and environmental sciences, as well as other fields related to its mission. NOAA has long conducted many types of education efforts, including grant making, creating partnerships with minority-serving colleges and universities, local hands-on learning experiences, and developing national science literacy frameworks. While its overall mission statement does not include education, dedication to education is underscored by the agency's vision to develop a society that is informed on issues related to its mission, and education objectives are implicit in each of the agency's four broad goals.

Conclusion I.2: *NOAA is unique among federal agencies in its focus on stewardship and on oceanic, coastal, Great Lakes, atmospheric, and climate sciences. However, its mission overlaps with and complements the missions of other federal agencies (Chapter 2: NOAA's Role in the Education Landscape).*

Several federal agencies share overlapping responsibilities for national programs in science, engineering, exploration, and stewardship related to oceans, the atmosphere, climate, and the environment. These agencies also develop and support programs in STEM education and environmental education that are similar to NOAA education programs. In addition, there is a significant infrastructure and human capacity outside the federal government for conducting scientific research, developing technology, and supporting education in areas related to NOAA's mission. Thus, NOAA needs to establish where its strengths lie, set priorities for how it engages in different education initiatives, and coordinate its efforts with those of other federal agencies.

Conclusion I.3: *NOAA has assumed a coordinating role on interagency groups convened to support coherence and collaboration across federal agencies involved in science education related to its mission (Chapter 1: Overview and Introduction; Chapter 4: Overview and Critique of NOAA's Education Programs).*

In the past decade, interagency coordination groups have been established to support coherence and collaboration through efforts, such as those to develop literacy frameworks that lay out the key ideas in oceanic, atmospheric, climate, and environmental sciences. NOAA is an active participant in these groups and has assumed leadership in some of them. For example, NOAA cochairs both the U.S. Climate Change Science Program's Education Interagency Working Group and the Ocean Research Resources Advisory Panel's Interagency Working Group on Ocean Education.

Conclusion I.4: *NOAA can fulfill its role in education by leveraging its modern and groundbreaking technologies and discoveries; research equipment; data; scientists, engineers, researchers, and other technical staff; stewardship and management of natural resources; specialized education expertise; partnerships; and connections to local, regional, national, and international stakeholders and natural resource managers (Chapter 2: NOAA's Role in the Education Landscape).*

NOAA is one of the key federal agencies engaged in stewardship of the coasts and oceans. The resources it manages provide vast and important education opportunities, and management of these environments provides the agency with connections to the surrounding communities and organizations concerned with environmental issues. NOAA brings to education the cutting-edge science, engineering, and technology (e.g., ocean research vehicles, satellite systems, data systems, data collecting buoys, etc.) that it produces through intramural activities or supports through funding of external scientists and engineers. Technology, scientific resources, and data systems provide opportunities for students and citizens to see scientists, engineers, and researchers in action; gain a scientific understanding of the natural world; and participate in research, data collection, and analysis.

Conclusion I.5: *NOAA's role in education is shaped by the distributed nature of its education efforts across the line offices and the Office of Education, small education staff, and small education budget (Chapter 2: NOAA's Role in the Education Landscape).*

Education programs are run by five of the agency's six line offices (the Office of Program Planning and Integration does not run education programs) and the Office of Education. The line offices (some of which have individual education mandates) and the Office of Education can act independently, sometimes even in competition with each other. Usually, an individual or a small team implements the majority of education programs at the local level. Most local NOAA staff can dedicate only a portion of their time to education, because many of them also have communication, extension, research, or exploration responsibilities. NOAA's education budget is also relatively small in comparison to other federal agencies engaged in STEM education, such as the U.S. Department of Education, the National Science Foundation, the National Aeronautics and Space Administration, and the U.S. Department of Energy. These limited resources make it next to impossible for NOAA to accomplish its ambitious education goals without strategic partnerships.

Conclusion I.6: *The scope of both the stewardship and science topics related to NOAA's mission is global in nature and involves partnerships at the local, state, federal, and international levels. NOAA can make more effective use of the agency's global and international programs in its domestic education activities and foster education and stewardship through collaborative work with international partners (Chapter 2: NOAA's Role in the Education Landscape).*

NOAA's mission includes topics that are inherently global, such as climate change and ocean observing, which makes it essential to establish and maintain strategic partnerships to accomplish its ambitious education goals. Clear education goals, planning, and strategic use of NOAA resources are critical aspects of effective partnerships.

Conclusion I.7: *NOAA can play a supporting role in state and local education. Its education efforts are more likely to be productive if they align with national and local education needs, because education activities and products that do not consider the needs of the potential audiences are less likely to be used (Chapter 2: NOAA's Role in the Education Landscape).*

Education programs need to focus on productive partnerships to support local and state education systems, while promoting NOAA's education and stewardship mission. As part of developing an implementation plan, NOAA will therefore need to make better use of assets and programs that already exist, such as academic research and education programs, sanctuaries and other protected areas managed by NOAA as well as other federal and state agencies, and even entertainment media, such as movies, radio, and computer and video games. In addition, NOAA can provide international, national, and state agencies with resources that clearly provide insight into the essential concepts and important principles of science and environmental education that should be woven into standards and curriculum expectations. NOAA's contribution to formal and informal education can be significant if the agency shares new research and makes this knowledge available and understood by the public.

Based on our conclusions regarding NOAA's role in education, the committee makes the following recommendations.

Recommendation I.1: NOAA should fulfill its role in education through the use of:

- agency and external expertise in science, engineering, technology, and education; cutting-edge scientific research and exploration activities; internationally collected data sets; and advances in technology and engineering;
- place-based assets that directly connect local issues to national and global science and stewardship issues: marine sanctuaries, estuarine research reserves, fisheries activities, and other natural resources protected and managed by federal, state, and local entities;
- partnerships with local and state education infrastructure, academic institutions, government agencies, business and industry, and private-sector and nonprofit organizations; and
- the agency's global science and international partnerships.

Recommendation I.2: In order to adequately address the mismatch between its available resources and its ambitious education agenda, NOAA should better align and deploy its resources. This may require the termination of certain activities and programs that, based on appropriate evaluation, do not directly and effectively contribute to its education and stewardship goals.

Recommendation I.3: Within the constraints of NOAA's mandates in education, the agency should continually evaluate where it leads, collaborates, follows, or declines to participate in partnerships with others. These decisions should be guided by consideration of the agency's role, assets, resources, and priorities in education and the strengths and missions of other agencies, institutions, and organizations engaged in education.

II. EDUCATION GOALS AND OUTCOMES

Two goals are outlined in the NOAA education strategic plan for 2009-2029: (1) to advance the environmental literacy of the nation and (2) to promote a diverse workforce in oceanic, coastal, Great Lakes, weather, and climate sciences. At this time, NOAA is developing a strategic implementation plan to specify how it will accomplish these goals. The strategic plan lists six outcomes under the environmental literacy goal:

1. The use of research on effective environmental and science education.
2. That educators understand and use environmental literacy principles.
3. That educators, students, and the public engage in inquiry-based learning.
4. That opportunities are available for lifelong engagement in science and environmental education.
5. That agencies collaborate on education activities.
6. That there is coordination across NOAA in education, extension, outreach, training, and communication.

Three outcomes are listed under the promoting a diverse workforce goal:

1. A diverse and qualified pool of applicants pursue opportunities for career development in NOAA mission-critical disciplines.
2. NOAA employees support activities to disseminate NOAA scientific and stewardship work.
3. A diverse pool of individuals who enter career paths at NOAA and in other related organizations.

The plan describes the issues and topics that need to be addressed to reach its goals and provides strategies to accomplish the outcomes under each goal. The plan also emphasizes the importance of partnerships to reach its education goals and outcomes.

Conclusion II.1: *The education strategic plan has a number of strengths:*

- *Appropriate goals of supporting environmental literacy and workforce development;*
- *A commitment to developing education programs informed by evidence about effective practices;*
- *A call to contribute to the body of knowledge on effective education practices in fields related to NOAA's mission;*
- *A call to develop partnerships with appropriate agencies, institutions, and organizations;*
- *An emphasis on including more members of historically underrepresented groups in fields critical to NOAA's mission; and*
- *An emphasis on the use of ocean, coastal, and other place-based resources as unique and valuable assets for learning (Chapter 3: The Education Portfolio and Effective Practices).*

The current education strategic plan is a step forward from the previous one. It provides more detailed description of outcomes and outlines strategies to bring about those outcomes. It also describes the importance of effective education practices, as well as issues and topics to be addressed in reaching the goals. Given the limited education resources of the agency, it is critical that the plan stresses the need to create partnerships and connections to local, regional, national, and international stakeholders and natural resource managers to achieve its education goals.

Conclusion II.2: *The current education strategic plan also has weaknesses that need to be addressed:*

- *Some of the nine outcomes are strategies for accomplishing the two overarching goals rather than measurable changes that would be expected as the result of reaching either goal.*
- *There is no environmental literacy outcome related to addressing the needs of underserved populations or addressing issues of equity.*
- *There is no environmental literacy outcome related to stewardship, yet many of the local education activities as implemented have stewardship goals.*
- *There is no clear articulation of the specific workforce needs, nor is there clear articulation or a plan to provide a robust assessment of the workforce needs in fields critical to NOAA's mission.*
- *There is no mechanism for local education staff from different education programs to share effective practices and lessons learned.*
- *There is no specific guidance or strategy for involving or drawing on NOAA expertise in engineering, science, and other fields.*
- *There is no specific guidance to NOAA employees engaged in education activities regarding fostering of external partners to connect to the national STEM infrastructure and human capacity (Chapter 3: The Education Portfolio and Effective Practices).*

These issues make it difficult to develop education programs that align the agency's education goals and objectives. Alignment would ensure that NOAA can capitalize on its assets. Alignment of goals and objectives would also support the agency's ability to conduct evaluations that produce data to show whether the agency is meeting its educational goals. These data are fundamental to decision making regarding the development of new programming and the management of existing programming.

To address these issues, NOAA needs to address three problems with the outcomes in the strategic education plan. First, outcomes are typically thought of as measurable changes or absolute levels of performance that can be expected as a result of the efforts made to reach a goal (i.e., what does it look like when environmental literacy is approached or reached and what constitutes a significant step along the path to acceptable diversity in the workforce). However, only three of the six environmental literacy outcomes and two of the three workforce development outcomes specify expectations; the other outcomes describe strategies or processes that might contribute to reaching the goals.

Second, while diversity is a focus of the workforce goal, there is no mention of diversity or broadening participation in the environmental literacy goal. There is a need to include an outcome focused on reaching out to underserved populations in the environmental literacy goal to address the national need to expand understanding of and interest in the science and stewardship issues related to NOAA's mission among K-12 students, adults, and the public.

Third, there is no environmental literacy outcome related to stewardship, despite the fact that stewardship is major element of environmental literacy. An outcome focused on stewardship is needed because it is the primary goal of many local education activities, and it would encourage the measurement of attitude and behavior change in program evaluations.

It is unclear how NOAA can accomplish its goal of supporting the creation of a "world-class" workforce without a clear understanding of its own and the nation's workforce needs in areas that are critical to its mission. Workforce needs are difficult to predict in general, and particularly difficult to do so in interdisciplinary areas, such as those critical to NOAA's mission. The agency needs a clearer estimate of workforce needs to guide the scope and direction of its initiatives.

Scientists and engineers have the expertise to introduce teachers and students to the processes of science and engineering, as well as to the cutting-edge research related to science and engineering activities that are connected to NOAA's mission. However, they need to work in concert with professionals who have specific expertise in learning, education, or behavior change and modification. The implementation of the education strategic plan or education implementation plan needs to provide guidance and require that these connections occur. Similarly, the implementation plan needs to provide guidance on fostering internal and external partnerships. Both are highlighted as important aspects of NOAA's education strategic plan, but there is very little detail regarding how either will be accomplished.

Conclusion II.3: *The use of the term "NOAA science" in the strategic plan is confusing. It is unclear whether this term is meant to refer to the science conducted by NOAA scientists, the research or the results of research funded by NOAA, or any science conducted on topics related to NOAA's mission (Chapter 3: The Education Portfolio and Effective Practices).*

It may be that the term "NOAA science" is used as a convenient shorthand term in describing broadly NOAA-related science. However, we are particularly concerned about narrow interpretations of the term as applied to NOAA's broader science education, stewardship, and workforce development goals. A narrow interpretation may lead to an exclusive focus on the research funded by NOAA, thereby constraining the agency's activities. Education activities aimed at environmental literacy for the public should draw on a broad body of scientific knowledge to avoid hampering public environmental literacy, which is a likely result of using a narrower body of knowledge. The committee is also concerned that the term may encourage a blurred line between activities focused on education and activities that are more appropriately defined as public relations or agency branding. This is not to say that there is evidence that the line is blurred in current activities, but rather that the use of the term "NOAA science" may lead to a blurring of this line in the future.

To address these concerns, the committee makes four recommendations.

Recommendation II.1: NOAA education programs should formally address broadening participation of underrepresented groups as an important outcome through all phases, from the initial stages of planning through implementation and evaluation. The environmental literacy goal, in particular, should include outcomes related to reaching out to underserved and underrepresented communities.

Recommendation II.2: To reach NOAA's environmental literacy goal, the Education Council should develop its implementation plan and future revisions of the education strategic plan to:

- clarify how it will capitalize on scientific findings, engineering advances, and stewardship activities that relate broad national priorities to local concerns to engage individuals of all ages in education;
- articulate how NOAA education programs will draw on the scientific, engineering, research, and other expertise accessible within the agency as well as in the broader community;
- address the mismatch between the lack of an outcome related to stewardship and the focus on stewardship outcomes in local programs;
- consistently define outcomes as measurable concepts that allow an assessment of whether a goal is being reached, to clearly distinguish outcomes on audiences (impact) from outputs of activities; and
- provide more opportunities for local and regional education staff from all education programs to share effective practices and lessons learned.

Recommendation II.3: To achieve the workforce development goal, the education strategic plan, the education implementation plan, or both should call for periodic assessment of the current and anticipated needs in fields critical to NOAA's mission to guide investment in appropriate workforce development activities.

Recommendation II.4: NOAA education programs should draw from current and relevant scientific and engineering advances regardless of what agency, institution, or organization they are originated or funded by.

III. COMPOSITION AND MANAGEMENT OF THE EDUCATION PORTFOLIO

NOAA supports a wide range of education programs for varied audiences that include K-12, postsecondary, graduate, and informal education activities with local, regional, national, and international scope. NOAA has developed professional development programs, classroom materials, curricula, museum exhibits, place-based learning experiences, literacy documents, and other products. The audiences for these programs include teachers, students, scientists, and the public. A coherent, coordinated education portfolio is needed to achieve goals effectively and efficiently, to share successful strategies for engaging and teaching different audiences, to pool resources to support synergistic activities, to develop cross-discipline activities, and to sustain consistent education strategies.

Conclusion III.1: *Management of a federal education portfolio is complicated, and NOAA has characteristics that make this particularly challenging:*

- *The fact that education programs are distributed across the agency.*
- *The overall management structure was not designed with education responsibilities in mind.*
- *A broad, overarching mission that includes a number of scientific areas, environmental stewardship, and commerce and transportation issues.*
- *Responsibilities that overlap and need to be coordinated with other federal agencies.*
- *The fact that development and implementation of nearly all of its education programs took place prior to the creation of the coordinating structures (the Education Council and the Education Office).*
- *The fact that some NOAA education programs have individual legislative mandates that guide their education activities (Chapter 2: NOAA's Role in the Education Landscape; Chapter 3: The Education Portfolio and Effective Practices).*

Creating and managing a strategically balanced education portfolio is no small task. Education programs are enacted by five of the six line offices in NOAA (the Office of Program Planning and Integration does not enact education programs) and the Office of Education. In addition, an unquantified amount of education support is provided through scientific research grants and contracts. Individual offices have separate mandates and often have local components with local control. Education programs are managed differently across these offices as a result of the resources available, separate missions, and different mandates. The differences in management structures, missions, and education mandates are obstacles to developing a cohesive and coordinated education portfolio. The Education Council needs to encourage the coordination of the agency's education programs to establish and monitor the agency's education portfolio.

Conclusion III.2: *While relatively new, the Education Council, led by the Office of Education, serves an essential, high-level internal coordinating function. The Education Council led the development of the education strategic plan and is developing the collaborative working relationships necessary for implementation. However, the Education Council does not have budgetary or institutional control over the education efforts of NOAA line and program offices; this limits its effectiveness in carrying out NOAA's education mandate (Chapter 3: The Education Portfolio and Effective Practices).*

The Education Council is the primary means that NOAA has developed to address the challenges in managing its education portfolio. In a short period of time, the Education Council has shown promise in bringing coherence and coordination to the agency's education portfolio by developing the strategic plan and encouraging a common evaluation framework. The committee is concerned that the Education Council may not have the needed power to enforce difficult decisions regarding the priorities, focus, and components in the portfolio in the near future—specifically, if the decisions require some line offices to end or restructure some of their education programs.

Providing more power to the Education Council does not come without concerns. It is not clear whether top-down decisions by the council will be perceived as supportive of the local education offices, which currently rely heavily on the enthusiasm, creativity, and good will of the local staff and their partners. Management structures or decisions that negatively impact the enthusiasm, creativity, and good will of local staff and partners could have far-reaching impacts on the success of the programs. Thus, any decisions about the power of the council must be weighed against the possibility of creating disharmony or a lack of enthusiasm through its decisions and decision-making process. In addition, the Education Council's decision-making process would be well served by honoring input from local staff of each of the line offices.

Conclusion III.3: *The NOAA education portfolio has developed in the absence of an overarching strategic direction and without a system to monitor or catalogue activities. It is therefore difficult to assess its composition, balance, and impact. In addition, there is no clear institutional mechanism to streamline or shift the portfolio (Chapter 4: Overview and Critique of NOAA's Education Programs).*

Although we received a great deal of information, the committee gained only a partial understanding of the education portfolio because the agency had no systematic process for collecting and cataloguing information about activities supported and carried out in the regional and local offices. It was therefore difficult to assess many aspects of the portfolio's balance. During our review, the agency began to develop such a process for collecting this information. Clearly, NOAA now understands that an adequate system to monitor and catalogue its education portfolio is needed to make informed decisions about portfolio balance.

The system to monitor and catalogue current and future portfolio balance should include evaluation criteria and evaluation procedures, continuing review of the entire portfolio to enable reprioritization, and new strategic directions. It should support the development of a dynamic portfolio in which programs may be discontinued if no longer needed or effective, to make room for new programs that may more effectively achieve goals and outcomes, or that allow NOAA to address new strategic goals and outcomes. Developing figures or tables, such as those presented in Chapter 4 (see Figure 4.2 and Table 4.2), could be useful in understanding and tracking portfolio balance.

Developing a coordinated and cohesive education portfolio is further complicated by the lack of a mechanism to bring the local education staff from different programs together to share effective practices across education programs. The strategic plan mentions the need for internal coordination to support education activities; however, the coordination being discussed is at the Education Council level. Promoting connections among local education staff can be just as valuable in creating internal coordination.

In addressing these concerns, NOAA will have to grapple with the level of centralized control that is optimal and information exchange systems that connect people who face similar issues. The agency must weigh whether it would be better off with a single, coordinated education program, or a set of loosely coupled programs run by the various offices, or whether a middle ground can be established. In making decisions about how to manage and monitor its education programs, NOAA should be striving to create an education portfolio that continues to address the needs of the communities it serves in an efficient and coordinated manner. In addition, the agency should have access to information needed to understand the composition of its programs, to make strategic decisions about the direction of its programs, and to evaluate the impact of its programs.

Conclusion III.4: *To date, NOAA's education programs have been more focused on ocean or coastal concepts and issues than on climate and atmospheric concepts and issues. There are emerging efforts to bring greater attention to climate and the atmosphere across the agency's education activities (Chapter 4: Overview and Critique of NOAA's Education Programs).*

Although we were unable to obtain information about all of NOAA's education programs, it is clear that the majority of activities focus on issues related to ocean literacy. The imbalance with respect to atmospheric and climate literacy is an issue acknowledged by NOAA education staff. The Climate Office, along with the Education Council, is working to bring greater attention to climate and atmospheric issues.

With regard to the composition and management of NOAA's education portfolio, the committee makes the following recommendation.

Recommendation III.1: NOAA should develop and implement a system to monitor and catalogue its education portfolio and guide decisions regarding what programs should be developed, continued, modified, or ended. In balancing the portfolio, the Education Council should

- increase attention to climate and atmospheric science education programs to complement the current focus on ocean science. These programs should emphasize the strong connections and interactions among the ocean, the atmosphere, the land, and human and nonhuman species;
- provide purposeful attention to both STEM learning and stewardship goals so as to enable synergies; and
- make decisions based on national education needs, the education priorities of the agency, and a clear picture of its education portfolio.

IV. EDUCATION EVALUATION PRACTICES

The challenges of carrying out appropriate evaluations of education projects are large. Most federal science agencies are struggling to meet these challenges, and NOAA is giving increasing attention to evaluating its education projects. Although evaluation is primarily the responsibility of each individual education and outreach program, the Education Council provides leadership and guidance for education evaluation activities across the agency. The strategic education plan 2009-2029 highlights the need for more comprehensive evaluation.

Conclusion IV.1: *NOAA has conducted evaluations of a small proportion of its educational activities, and these evaluations are limited in scope and quality. The evaluations tend to focus on intermediate rather than long-term outcomes and on participant opinion, feedback, beliefs, and knowledge. They usually do not address outcomes related to attitudes or behavior, and they generally lack control or comparison groups. Summative evaluations have been carried out on a very small proportion of education activities, and there has been little consideration of evaluation that would allow NOAA to recalibrate the education portfolio to effectively meet the agency's educational goals (Chapter 4: Overview and Critique of NOAA's Education Programs; Chapter 5: Current Evaluation Framework and Existing Evaluation Efforts).*

While NOAA education programs have conducted many evaluations, they have resulted in very few data that illustrate the impact of the programs. For example, many evaluations include only self-reported information about participant enjoyment, satisfaction, or perceived impact. Also, even when appropriate there were rarely control or comparison groups in the evaluations we reviewed. Evaluation has historically not been used by the agency to collect data to assess whether the overarching goals and outcomes in the education strategic plan are being accomplished.

One major issue is that evaluation, particularly when focusing on impact at the program level, can be expensive. A rule of thumb is that at least 5 percent of a projects budget should be devoted to summative evaluation. Formative evaluation should be part of program design, and its cost is part of the program. However, the education efforts in NOAA are operating on a shoestring budget, and taking 5 percent of the budget away from a project's operating costs may negatively impact the project's implementation and be insufficient to conduct a meaningful evaluation.

Given its limited overall funds, it is critical that NOAA develop a plan for wisely allocating funds for evaluation (so as not to compromise programming). To achieve the greatest return on its limited resources, evaluations of individual projects can be scheduled on a cyclical basis, with high priority given to projects intended to have the greatest impact on environmental literacy and workforce needs and to projects that face important questions regarding activities, participants, staffing, funding, or organization. In addition, partnerships can be developed to minimize the financial burden of the evaluation process on NOAA. Program effectiveness should be determined by conducting formative evaluations as part of the project or program design process.

Conclusion IV.2: *The Education Council is increasing its emphasis on evaluation and moving toward comprehensive program evaluation by adopting the Bennett Targeting Outcomes of Programs (TOP) model. This approach may help implement a more strategic and coherent approach to education evaluation efforts across the agency, serving as an important tool to align programs with NOAA's education goals. However, as with most evaluation models, the TOP model does not include specific guidance regarding the implementation of evaluations or how to design high-quality evaluations (Chapter 5: Current Evaluation Framework and Existing Evaluation Efforts).*

The adoption of a uniform model, such as the Bennett TOP model, as a framework to guide evaluation strategies and practices is a useful step that illustrates the positive influence of the Education Council. NOAA education programs need to continue to refine their evaluation expertise, contract with appropriate evaluators to design and implement evaluations, and use evaluation results to improve education activities.

Conclusion IV.3: *The collection of data for project monitoring and evaluation purposes is uneven across the education portfolio. A set of guidelines to systematically collect consistent data for these purposes is needed (Chapter 4: Overview and Critique of NOAA's Education Programs; Chapter 5: Current Evaluation Framework and Existing Evaluation Efforts).*

Data are needed for several purposes, including project monitoring, fiscal due diligence, and program evaluation. NOAA needs a systematic way of collecting data for each purpose to ensure that data are comparable across programs and initiatives.

Education evaluation practices were being augmented during our review of NOAA's education program. To ensure that these practices continue to develop in a positive direction, the committee makes the following recommendations.

Recommendation IV.1: The Education Council should continue to improve the evaluation expertise of its education program managers, contract with external evaluators for summative evaluation, and require the incorporation of the most appropriate and rigorous evaluation strategies during program development to guide design, continual improvement, and delivery of its education programs.

Recommendation IV.2: The Education Council should increase the emphasis on high-quality evaluations. Summative evaluations should focus on the program outcomes related to learning and stewardship, not only satisfaction with education experiences, and should use the most appropriate and rigorous evaluation designs.

Recommendation IV.3: The Education Council should consider developing a number of approaches to inform strategic portfolio management and how evaluation findings can be used to inform decisions about portfolio balance.

Recommendation IV.4: Education programs should evaluate internal collaboration among line offices and between education and operational and scientific staff, as well as the quality of external partnerships with other agencies, institutions, organizations, and the broader STEM communities.

V. EVIDENCE OF IMPACT

Many of the education programs have not been in existence long enough to allow any definitive impact assessments. Furthermore, because of the problems with the outcomes defined in the strategic plan and the lack of rigorous evaluation programs, the Education Council has not yet developed the capability to analyze the impact of the larger portfolio of education programs. In general, there is little comparative, long-term, or causal data about individual programs or across programs to enable such analyses.

Conclusion V.1: *Although NOAA has created a large number of education initiatives with its limited education budget, there is evidence of impact for only a small proportion of them. The majority of initiatives have collected information only on their scope, reach, and participant satisfaction, which are not sufficient to judge impact. On the basis of the available evidence, all that can be said about that the impact of such programs is that they are positively perceived by the participants and, as yet, serve a small portion of the population (Chapter 4: Overview and Critique of NOAA's Education Programs).*

There is limited information regarding the impact of NOAA's education programs—a situation that is not uncommon among federal agencies. However, NOAA's ability to make good use of its education resources and assets to engage in a substantial number of education activities is impressive. Most partnerships with educational organizations, other agencies, and institutions with complementary STEM expertise have enhanced the reach and impact of NOAA's education efforts. These partnerships have often yielded additional expertise, educational tools, mechanisms for dissemination, and matching funds.

Conclusion V.2: *Although the current strategic education plan calls for the use of research-based education practices, current education activities do not consistently follow what is understood about effective education practices in the United States and abroad (Chapter 3: The Education Port-*

folio and Effective Practices; Chapter 4: Overview and Critique of NOAA's Education Programs).

There is a growing body of literature regarding effective practices in formal and informal science education, behavior change, reaching underserved populations, and workforce preparation. This literature can be used to support the development of science education programs that are likely to be successful.

To encourage the development and implementation of effective programs to address the goals outlined in NOAA's education strategic plan, the committee makes the following recommendation.

Recommendation V.1: NOAA education staff should draw on evidence from education research, evaluations of NOAA programs, and external education expertise to identify and implement effective practices for supporting education activities.

CONCLUDING NOTE

People are NOAA's most valuable assets. The education staff is dedicated and passionate about addressing areas related to its mission. They have developed diverse education activities for a wide range of audiences and regions. While many of the conclusions of this committee address issues with NOAA's education efforts, the agency and its education staff are to be commended for their historic commitment to education, which precedes the agencywide congressional mandate on education. The agency's current education strategic plan is a significant improvement over the previous one. We hope that our recommendations help NOAA continue to improve its education efforts.

References

American Association for the Advancement of Science. (1989). *Report of the Project 2061 Phase 1*. Washington, DC: Author.

American Association for the Advancement of Science. (1993). *Benchmarks for science literacy*. New York: Oxford University Press.

American Association for the Advancement of Science. (2009). *R&D budget and policy program FY 2010 appropriations: Department of Education*. Available: http://www.aaas.org/spp/rd/09ptbii18.pdf [accessed November 2009].

American Geological Institute. (2009). *Status of the geoscience workforce 2009 edition*. Washington, DC: Author.

Association of American Colleges and Universities. (2006). *National defense education and innovation initiative: Meeting America's economic and security challenges in the 21st century*. Washington, DC: Author.

Association of American Colleges and Universities. (2007). *College learning for the new global century: A report from the National Leadership Council for Liberal Education and America's Promise*. Available: http://www.aacu.org/advocacy/leap/documents/GlobalCentury_final.pdf [accessed May 2007].

Ball, D.L., and Cohen, D.K. (1999). Developing practice, developing practitioners: Toward a practice-based theory of professional education. In G. Sykes and L. Darling-Hammond (Eds.), *Teaching as the learning profession: Handbook of policy and practice* (pp. 3-32). San Francisco: Jossey Bass.

Bang, M., Medin, D.L., and Atran S. (2007). Cultural mosaics and mental models of nature. *Proceedings of the National Academy of Sciences of the United States of America, 104*(35), 13868-13874.

Bardach, E., and Lesser, C. (1996). Accountability in human services collaboration: For what? And to whom? *Journal of Public Administration Research and Theory, 6*(2), 197-224.

Barnett, C. (1998). Mathematics teaching cases as a catalyst for informed strategic inquiry. *Teaching and Teacher Education, 14*(1), 81-93.

Belden, L., Blaustein A., Olson, D.H., Green, D.H., Root, T.L., and Kiesecker, J.M. (2001). Amphibian breeding and climate change. *Conservation Biology, 115*, 1804-1809.

Bennett, C. (1975). Up the hierarchy. *Journal of Extension, 13*(2), 7-12.

Bennett, C. (1979). *Analyzing impacts of extension programs.* Science and Education Administration (#ESC-57). Washington, DC: U.S. Department of Agriculture.

Bennett, C., and Rockwell, K. (1995). *Targeting outcomes of programs (TOP): An integrated approach to planning and evaluation.* Available: http://citnews.unl.edu/TOP/english/index.html [accessed April 2010].

Bogner, F.X. (1998). The influence of short-term outdoor ecology education on long-term variables of environmental perspective. *Journal of Environmental Education, 29*(4), 17-29.

Bord, S., Horner, A., Beavan, S., and Compston, J. (2001). Estrogen receptors α and β are differentially expressed in developing human bone. *Journal of Clinical Endocrinology and Metabolism, 86*, 2309-2314.

Brackett, A. (2009). *Review of NOAA education program evaluation reports.* Paper prepared for the Committee for the Review of the NOAA Education Program, National Research Council, Washington, DC. Available: http://www7.nationalacademies.org/bose/NOAA%20Evaluation.pdf [accessed March 2010].

Brinkerhoff, J.M. (2002). *Partnership for international development: Rhetoric or results.* Boulder, CO: Lynne Rienner.

Building Engineering and Science Talent. (2004). *A bridge for all: Higher education design principles to broaden participation in science, technology, engineering and mathematics.* San Diego: Author. Available: http://www.bestworkforce.org/PDFdocs/BEST_BridgeforAll_HighEdFINAL.pdf [accessed May 2007].

Business Roundtable. (2005, July 16). *Tapping America's potential: The education for innovation initiative.* Washington, DC: Author.

Cammen, L. (2008). *NOAA Sea Grant education: Evaluation overview.* Available: http://www.seagrant.noaa.gov/other/admininfo/presentations/NRCpresentationLeonCammen.pdf [accessed October 2009].

Carey, S. (1985). *Conceptual change in childhood.* Cambridge, MA: MIT Press.

Carnegie-IAS Commission on Mathematic and Science Education. (2009). *The opportunity equation: Transforming mathematics and science education for citizenship and the global economy.* Available: http://www.opportunityequation.org/ [accessed October 2009].

Center for Children and Technology. (2003). *JASON multimedia science curriculum's impact on student learning: Final evaluation report, year three.* New York: Author.

Center on Education Policy. (2007). *Choices, changes, and challenges: Curriculum and instruction in the NCLB era.* Washington, DC: Author.

Clune, B. (2009). *Logic models of NOAA education programs and implications for restructuring.* Paper prepared for the Committee for the Review of the NOAA Education Program, National Research Council, Washington, DC. Available: http://www7.nationalacademies.org/bose/NOAA%20Logic%20Model.pdf [accessed March 2010].

Coley, J.D., Vitkin, A.Z., Seaton, C.E., and Yopchick, J.E. (2005). Effects of experience on relational inferences in children: The case of folk biology. In B.G. Bara, L. Barsalou, and M. Bucciarelli (Eds.), *Proceedings of the 27th Annual Conference of the Cognitive Science Society* (pp. 471-475). Mahwah, NJ: Lawrence Erlbaum Associates.

Congressional Research Service. (2008). *Science, technology, engineering, and mathematics (STEM) education: Background, federal policy, and legislative action.* Available: http://assets.opencrs.com/rpts/RL33434_20080321.pdf [accessed October 2009].

Coral Reef Conservation Program. (2007). *Coral Reef Conservation Program 2002-2006 External review: Final report.* Available: http://coralreef.noaa.gov/aboutcrcp/strategy/reprioritization/exreview/resources/summary_report.pdf [accessed March 2009].

Cronbach, L. (1982). *Designing evaluations of educational and social programs.* San Francisco: Jossey-Bass.

Cucker, B. (2001). Steps to increasing minority participation in the aquatic sciences: Catching up with shifting demographics. *American Society of Limnology and Oceanography Bulletin, 10*, 17-21.

Day-Miller, E., and Payne, D. (2008). *Final evaluation report for the NOAA Ocean Exploration and Research Professional Development Institutes*. Unpublished report.

Dickerson, D.L., and Dawkins, K.R. (2004). Use of scientific fieldwork in designing high-quality professional development programs. *Teacher Education Journal of South Carolina*, 52-56.

Dillon, J., Rickinson, M., Teamey, K., Morris, M., Choi, M.Y., Sanders, D., and Benefield, P. (2006). The value of outdoor learning: Evidence from research in the UK and elsewhere. *School Science Review, 87*, 107-111.

Dorph, R., Goldstein, D., Lee, S., Lepori, K., Schneider, S., and Venkatesan, S. (2007). *The status of science education in the Bay Area: Research brief*. Berkeley: Lawrence Hall of Science, University of California.

Eberbach, C., and Crowley, K. (2005). From living to virtual: Learning from museum objects. *Curator, 48*(3), 317-338.

Education Trust. (2008). *Core problems: Out-of-field teaching persists in key academic courses and high-poverty schools*. Washington, DC: Author.

Edwards, C.H. (1997). Promoting student inquiry. *Science Teacher, 64*(7), 18-21.

Falk, J.H., and Dierking, L.D. (2000). *Learning from museums: Visitor experiences and the making of meaning*. Walnut Creek, CA: AltaMira Press.

Federal Coordinating Council for Science, Engineering, and Technology. (1992). *Policies and procedures manual*. Available: http://www.eric.ed.gov/ERICDocs/data/ericdocs2sql/content_storage_01/0000019b/80/15/41/fc.pdf [accessed October 2009].

Fortus, D., Hug, B., Krajcik, J., Kuhn, L., McNeill, K., Reiser, B., Rivet, A., Rogat, A., Schwartz, C., and Shwartz, Y. (2006). *Sequencing and supporting complex scientific inquiry practices in instructional materials for middle school students*. Paper presented at the annual meeting of the National Association for Research in Science Teaching, San Francisco, April. Available: http://www.umich.edu/~hiceweb/iqwst/Papers/NARST2006_Learning_progress.pdf [accessed February 2010].

Fox Keller, E. (1995). *Reflections on gender and science: 10th anniversary edition*. New Haven, CT: Yale University Press.

Garet, M., Porter, A., Desimone, L., Birman, B., and Kwang, S. (2001). What makes professional development effective? Results from a national sample of teachers. *American Educational Research Journal, 38*(4), 915-945.

Gentner, D., and Colhoun, J. (2008). Analogical processes in human thinking and learning. In B. Glatzeder, V. Goel, and A.V. Muller (Eds.), *On thinking: Volume 2. Towards a theory of thinking*. Berlin: Springer-Verlag.

Gentner, D., and Toupin, C. (1986). Systematicity and surface similarity in the development of analogy. *Cognitive Science, 10*, 277-300.

Gerhart, M., Saxe, G.B., Seltzer, M., Schlackman, J., Ching, C.C., Nasir, N.N., Fall, R., Bennett, T., Rhine, S., and Slone, T.F. (1999). Opportunities to learn fractions in elementary mathematics classrooms. *Journal for Research in Mathematics Education, 30*(3), 286-315.

Gilligan, M.G., Verity, P.G., Cook, C.B., Cook, S.B., Booth, M.G., and Frischer, M.E. (2007). Building a diverse and innovative ocean workforce through collaboration and partnerships that integrate research and education: HBCUs and marine laboratories. *Journal of Geoscience Education, 55*, 531-540.

Goodlad, J.I., and Sirotnik, K.A. (1988). The future of school-university partnerships. In K.A. Sirotnik and J.I. Goodlad (Eds.), *School-university partnerships in action* (pp. 205-225). New York: Teachers College Press.

Griffin, J. (1998). *School-museum integrated learning experiences in science: A learning journey.* Unpublished doctoral dissertation, University of Technology, Sydney.

Griffith, G., and Scharmann, L. (2008). Initial impacts of No Child Left Behind on elementary science education. *Journal of Elementary Science Education, 20*(3), 35-48.

Guskey, T.R. (2000). *Evaluating professional development.* Thousand Oaks, CA: Corwin Press.

Hammerman, J.K.L. (2007). *The state of estuaries education: K-12 needs assessment.* Cambridge, MA: TERC.

Hassett, S., and Austin, M.J. (1997). Service integration: Something old and something new. *Administration in Social Work, 21*, 9-29.

Hewson, P.W. (2007). Teacher professional development in science. In S.K. Abell and N.G. Lederman (Eds.), *Handbook of research on science education* (pp. 1179-1203). Mahwah, NJ: Lawrence Erlbaum Associates.

Hill, C.J., and Lynn, L.E. (2003). Producing human services: Why do agencies collaborate? *Public Management Review, 5*(1), 63-81.

Hmelo-Silver, C.E., Marathe, S., and Liu, L. (2007). Fish swim, rocks sit, and lungs breathe: Expert-novice understanding of complex systems. *Journal of Learning Sciences, 16*, 307-331.

Hoffman, M., and Barstow, D. (2007). *Revolutionizing earth system science education for the 21st century.* (Report and recommendations from a 50-state analysis of earth science education standards). Cambridge, MA: TERC.

Huffman, D., Thomas, K., and Lawrenz, F. (2003). Relationship between professional development, teachers' instructional practices, and the achievement of students in science and mathematics. *School Science and Mathematics, 103*(8), 378-387.

Huntoon, J., and Lane, M. (2007). Diversity in the geosciences and successful strategies for increasing diversity. *Journal of Geoscience Education, 55*(6), 447-457.

Inagaki, K. (1990). The effects of raising animals on children's biological knowledge. *British Journal of Developmental Psychology, 8*, 119-129.

Johnson, P. (1998). Children's understanding of changes of state involving the gas state, Part 1: Boiling water and the particle theory. *International Journal of Science Education, 20*(5), 567-583.

Joint Committee on Standards for Educational Evaluation. (1994). *Program evaluation standards.* Thousand Oaks, CA: Corwin Press.

Jones, M.G., Taylor, A., Minogue, J., Broadwell, B., Wiebe, E., and Carter, G. (2007a). Understanding scale: Powers of ten. *Journal of Science Education and Technology, 16*(2), 1059-1145.

Jones, M.G., Tretter, T., Taylor, A., and Oppewal, T. (2007b). Experienced and novice teachers' concepts of spatial scale. *International Journal of Science Education, 30*(3), 409-429.

Kastens, K.A., and Rivet, A. (2008, January). Multiple modes of inquiry in earth science. *Science Teacher*, 26-31.

Kastens, K.A., and Turrin, M. (2008). What are children being taught in school about anthropogenic climate change? In B. Ward and S. Mendez (Eds.), *Communicating on climate change: An essential resource for journalists, scientists and educators* (pp. 48-49). Kingston: University of Rhode Island.

Kastens, K.A., Manduca, C.A., Cervato, C., Frodeman, R., Goodwin, C., Liben, L.S., Mogk, D.W., Spangler, T.C., Stillings, N.A., and Titus, S. (2009). How geoscientists think and learn. *EOS, Transactions of the American Geophysical Union, 90*(31), 265-266.

Kauffman Foundation. (2007). *Education, entrepreneurship and immigration: America's new immigrant entrepreneurs, Part II.* Available: http://www.kauffman.org/uploadedFiles/entrep_immigrants_1_070907.pdf [accessed October 2009].

Klahr, D., Triona, L., and Siler, S. (2008). Virtual versus physical materials in early science instruction: Transitioning to an autonomous tutor for experimental design. In J. Zumbach, N. Schwartz, T. Seufert, and L. Kester (Eds.), *Beyond knowledge: The legacy of competence* (pp. 163-172). Dordrecht: Springer Netherlands.

Klahr, D., Triona, L.M., and Williams, C. (2007). Hands on what? The relative effectiveness of physical vs. virtual materials in an engineering design project by middle school children. *Journal of Research in Science Teaching, 44*, 183-203.

Kollmuss, A., and Agyeman, J. (2002). Mind the gap: Why do people act environmentally and what are the barriers to pro-environmental behavior? *Environmental Education Research, 8*(3), 239-260.

Leinhardt, G., and Crowley, K. (2002). Objects of learning, objects of talk: Changing minds in museums. In S. Paris (Ed.), *Multiple perspectives on object-centered learning* (pp. 301-324). Mahwah, NJ: Lawrence Erlbaum Associates.

Leiserowitz, A. (2006). Climate change risk perception and policy preferences: The role of affect, imagery, and values. *Climatic Change, 77*, 1-2.

Lemke, M., and Gonzales, P. (2006). *U.S. student and adult performance on international assessments of educational achievement: Findings from The Condition of Education 2006.* (NCES #2006-073). Washington, DC: U.S. Department of Education, National Center for Education Statistics.

Levine, R., González, R., and Martínez-Sussmann, C. (2009). *Learning diversity in earth system science.* Paper prepared for the Committee for the Review of the NOAA Education Program, National Research Council, Washington, DC. Available: http://www7.nationalacademies.org/bose/NOAA%20Diversity.pdf [accessed March 2010].

Lewis, J.A. (2005). *Waiting for Sputnik: Basic research and strategic competition.* Washington, DC: Center for Strategic and International Studies. Available: http://csis.org/publication/waiting-sputnik [accessed March 2010].

Longino, H.E. (2003). Does the structure of scientific revolutions permit a feminist revolution in science? In T. Nickles (Ed.), *Thomas Kuhn (contemporary philosophy in focus).* Cambridge, MA: Cambridge University Press.

Loucks-Horsley, S., Hewson, P.W., Love, N. and Stiles, K.E. (1998). *Designing professional development for teachers of science and mathematics.* Thousand Oaks, CA: Corwin Press.

Loucks-Horsley, S., Love, N., Stiles, K.E., Mundry, S., and Hewson, P.W. (2003). *Designing professional development for teachers of science and mathematics* (2nd ed.). Thousand Oaks, CA: Corwin Press.

Louis, R.P. (2007). Can you hear us now? Voices from the margin: Using indigenous methodologies in geographic research. *Geographical Research, 45*(2), 130-139.

Lovelace, G. (2006). *Evaluation of the benefits to scientists of participation in outreach and education projects related to Ocean Explorations Expeditions.* Unpublished evaluation report.

Mason, L., and Santi, M. (1998). Discussing the greenhouse effect: Children's collaborative discourse-reasoning and conceptual change. *Environmental Education Research, 4*(1), 67-85.

McBeth, B., Hungerford, H., Marcinkowski, T., Volk, T. and Meyers, R. (2008) *National Environmental Literacy Assessment Project: Year 1, national baseline study of middle grades students.* Final research report submitted to the Environmental Education Division of the Environmental Protection Agency and Office of Education of the National Oceanic and Atmospheric Administration.

McKinsey and Company. (2009). *The economic impact of the achievement gap in American schools.* Available: http://www.mckinsey.com/App_Media/Images/Page_Images/Offices/SocialSector/PDF/achievement_gap_report.pdf [accessed October 2009].

Mertens, D., and Hopson, R. (2006). Advancing evaluation of STEM efforts through attention to diversity and culture. *New Directions for Evaluation, 109*, 35-51.

Mewborn, D.S. (2003). Teaching, teachers, knowledge, and their professional development. In J. Kilpatrick, W.G. Martin, and D. Schifter (Eds.), *A research companion to principals and standards for school mathematics* (pp. 45-52). Reston, VA: National Council of Teachers of Mathematics.

Moore, B., Rosen, R.D., Rosenberg, A.A., Spinrad, R.W., Washington, W.M., and West, R.D. (2004). *Review of the organization and management of research in NOAA: A report to the NOAA science advisory board.* Available: http://www.sab.noaa.gov/Reports/RRT_Report-080604.pdf [accessed October 2009].

Mullis, I.V.S., Martin, M.O., Foy, P., Olson, J.F., Preuschoff, C., Erberber, E., Arora, A., and Galia, J. (2007). *Trends in International Mathematics and Science Study (TIMSS) 2007.* Chestnut Hill, MA: Boston College.

National Academy of Engineering. (2005). *Educating the engineer of 2020: Adapting engineering education to the new century.* Committee on the Engineer of 2020, Phase II. Committee on Engineering Education, Washington, DC: The National Academies Press.

National Academy of Engineering and National Research Council. (2009). *Engineering in K-12 education: Understanding the status and improving the prospects.* Committee on K-12 Engineering Education. L. Katehi, G. Pearson, and M. Feder (Eds.). Washington, DC: The National Academies Press.

National Academy of Sciences. (1929). *Report of the Committee of Oceanography for the National Academy of Sciences.* Washington, DC: National Academy Press.

National Academy of Sciences, National Academy of Engineering, and Institute of Medicine. (2000). *Enhancing the postdoctoral experience for scientists and engineers: A guide for postdoctoral scholars, advisers, institutions, funding organizations, and disciplinary societies.* Committee on Science, Engineering, and Public Policy. Washington, DC: National Academy Press.

National Academy of Sciences, National Academy of Engineering, and Institute of Medicine. (2007). *Rising above the gathering storm: Energizing and employing America for a brighter economic future.* Committee on Prospering in the Global Economy of the 21st Century: An Agenda for American Science and Technology. Committee on Science, Engineering, and Public Policy. Washington, DC: The National Academies Press.

National Center on Education and the Economy. (2006). *Tough choices for tough times: The report of the New Commission on the Skills of the American Workforce.* Washington, DC: Author.

National Oceanic and Atmospheric Administration. (2009). *Education strategic plan 2009-2029.* Washington, DC: Author.

National Oceanic and Atmospheric Administration, Science Advisory Board. (2008). *Engaging NOAA's constituents, a report from the NOAA Science Advisory Board: Putting the pieces together to create impacts.* Available: http://www.sab.noaa.gov/Reports/EOEWG/EOEWG_Final_Report_03_20_08.pdf [accessed February 2009].

National Research Council. (1996). *National science education standards.* National Committee on Science Education Standards and Assessment Center for Science, Mathematics, and Engineering Education. Commission on Behavioral and Social Sciences and Education. Washington, DC: National Academy Press.

National Research Council. (1999). *How people learn: Brain, mind, experience, and school.* J.D. Bransford, A.L. Brown, and R.R. Cocking (Eds.). Committee on Developments in the Science of Learning. Commission on Behavioral and Social Sciences and Education. Washington, DC: National Academy Press.

National Research Council. (2002). *New tools for environmental protection: Education, information, and voluntary measures.* Committee on the Human Dimensions of Global Change. T. Dietz and P.C. Stern (Eds.). Division of Behavioral and Social Sciences and Education. Washington, DC: The National Academies Press.

National Research Council. (2005). *Assessment of NIH minority research and training programs: Phase 3.* Committee for the Assessment of NIH Minority Research Training Programs, Oversight Committee for the Assessment of NIH Minority Research Training Programs. Board on Higher Education and Workforce. Washington, DC: The National Academies Press.

National Research Council. (2006). *Issues affecting the future of the U.S. space science and engineering workforce.* Committee on Meeting the Workforce Needs for the National Vision for Space Exploration, Division on Engineering and Physical Sciences. Washington, DC: The National Academies Press.

National Research Council. (2007). *Taking science to school.* Committee on Science Learning, Kindergarten Through Eighth Grade. R.A. Duschl, H.A. Schweingruber, and A.W. Shouse (Eds.). Board on Science Education, Center for Education. Division of Behavioral and Social Sciences and Education. Washington, DC: The National Academies Press.

National Research Council. (2008). *NASA's elementary and secondary education program: Review and critique.* Committee for the Review and Evaluation of NASA's Precollege Education Program. H.R. Quinn, H.A. Schweingruber, and M.A. Feder (Eds.). Board on Science Education, Center for Education. Division of Behavioral and Social Sciences and Education. Washington, DC: The National Academies Press.

National Research Council. (2009). *Learning science in informal environments.* Committee on Learning Science in Informal Environments. P. Bell, B. Lewenstein, A.W. Shouse, and M.A. Feder (Eds.). Board on Science, Center for Education. Division of Behavioral and Social Sciences and Education. Washington, DC: The National Academies Press.

National Science Board. (2003). *The science and engineering workforce: Realizing America's potential.* Arlington, VA: National Science Foundation.

National Science Board. (2004). *Broadening participation in science and engineering faculty.* Arlington, VA: National Science Foundation.

National Science Board. (2007). *A national action plan for addressing the critical needs of the U.S science, technology, engineering, and mathematics education system.* Available: http://www.nsf.gov/pubs/2007/nsb07114/nsb07114.pdf [accessed October 2009].

National Science Board. (2009). *Science and engineering indicators: 2008.* Arlington, VA: National Science Foundation.

National Science Foundation. (1996). *Women, minorities, and persons with disabilities in science and engineering: 1996.* Arlington, VA. Author.

National Science Foundation. (2005). *Federal obligations for basic research, by agency and performer: FY 2005.* Available: http://www.nsf.gov/statistics/nsf09300/pdf/tab27.pdf [accessed October 2009].

National Science Foundation. (2010). *Science and engineering degrees, by race/ethnicity of recipients: 1997–2006, detailed statistical tables.* (NSF #10-300). Arlington, VA. Available: http://www.nsf.gov/statistics/nsf10300/ [accessed October 2009].

National Summit on Competitiveness. (2005). *Statement of the National Summit on Competitiveness: Investing in U.S. innovation.* Washington, DC: Author.

North American Association for Environmental Education. (2004). *Guidelines for the preparation and professional development of environmental educators.* Available: http://www.naaee.org/npeee/new_ee.php [accessed February 2009].

Ocean Explorer. (2002). *The sea with no shores.* Available: http://oceanexplorer.noaa.gov/explorations/02sab/background/edu/media/sab_sargassum.pdf [accessed October 2009].

Office of National Marine Sanctuaries. (2004). *Dive into education program: Executive summary and evaluation.* Available: http://sanctuaries.noaa.gov/education/teachers/features/diveeducation.html [accessed October 2009].

Office of National Marine Sanctuaries. (2005a). *West Coast LiMPETS teacher post-workshop evaluation.* Unpublished evaluation report, Silver Spring, MD.

Office of National Marine Sanctuaries. (2005b). *National Marine Sanctuary Program Hawaii field study 2005.* Unpublished evaluation report, Silver Spring, MD.

Pandion Systems. (2003). *Inventory and assessment of K-12 and professional teacher development programs in NOAA's national estuarine research reserve system.* Available: http://www.estuaries.gov/estuaries101/Doc/PDF/InventoryK12Report.pdf [accessed October 2009].

Pandion Systems. (2005). *Estuaries live: Final report.* Available: http://cfpub.epa.gov/ncer_abstracts/INDEX.cfm/fuseaction/display.abstractDetail/abstract/6118/report/F [accessed October 2009].

Pandion Systems. (2006). *Estuaries live: Final report.* Available: http://steinhardt.nyu.edu/steinhardt/wallerstein/pdf_uploads/Online%20Hudson%20River%20Report%202005.pdf [accessed October 2009].

Pew Ocean Commission. (2003). *America's living oceans, charting a course for sea change.* Available: http://www.pewtrusts.org/uploadedFiles/wwwpewtrustsorg/Reports/Protecting_ocean_life/env_pew_oceans_final_report.pdf [accessed October 2009].

Plumb, C., and Reis, R.M. (2007). Creating change in engineering education: A model of collaboration among institutions. *Change, the Magazine of Higher Learning, 39*(3), 22-29.

Project Kaleidoscope. (2006). *Transforming America's scientific and technological infrastructure: Recommendations for urgent action. Report on reports II.* Available: http://www.pkal.org/documents/ReportOnReportsII.cfm [accessed October 2009].

Rockwell, K., and Bennett, C. (2004). *Targeting outcomes of programs: A hierarchy for targeting outcomes and evaluating their achievement.* Available: http://digitalcommons.unl.edu/aglecfacpub/48 [accessed October 2009].

Scherer, J. (2008). A review of instruments to evaluate partnerships in math and science education. *Journal of Education, 83*(4), 611-636.

Schifter, D. (1996). *What's happening in math class? Envisioning new practices through teacher narratives* (vols. 1-2). New York: Teachers College Press.

Seymour, E., and Hewitt, N.M. (1997). *Talking about leaving: Why undergraduates leave the sciences.* Boulder, CO: Westview Press.

Snively, G. (1995). Bridging traditional science and western science in the multicultural classroom. In G. Snively and A. MacKinnon (Eds.), *Thinking globally about mathematics and science education* (pp. 53-75). Vancouver: University of British Columbia.

State Higher Education Executive Officers. (2008). *State higher education finance FY 2008.* Available at http://www.sheeo.org/finance/shef_fy08.pdf [accessed October 2009].

Steel, B., Lovrich, N., Lach, D., and Fomenko, V. (2005) Correlates and consequences of public knowledge concerning ocean fisheries management. *Coastal Management, 33*(1), 37-51.

Steel, B., Smith, C., Opsommer, L., Curiel, S., and Steel, R.W. (2005). Public ocean literacy in the United States. *Ocean and Coastal Management, 48*, 97-114.

Stufflebeam, D. (2005) CIPP model (context, input, process, product). In S. Mathison (Ed.), *International handbook of educational evaluation* (pp. 60-65). Thousand Oaks, CA: Sage.

Stufflebeam, D., and Shinkfield, A. (2007). *Evaluation theory, models, and applications.* San Francisco: Jossey-Bass.

Sullivan, D., Rosenfeld, L., and Murphree, T. (2007, winter). Certification for oceanographic professionals: A needs assessment study. *Orion Newsletter, 5.*

Swanson, C.B. (2009). *Cities in crisis 2009: Closing the graduation gap*. Bethesda, MD: Editorial Projects in Education.

Swim, J., Clayton, S., Doherty, T., Gifford, R., Howard, G., Reser, J., Stern, P., and Weber, E. (2009). *Psychology and global climate change: Addressing a multifaceted phenomenon and set of challenges.* Report of the American Psychological Association Task Force on the Interface Between Psychology and Global Climate Change. Available: http://www.apa.org/science/climate-change [accessed October 2009].

Tai, R.H., Liu, C.Q., Maltese, A.V., and Fan, X. (2006). Planning early careers in science. *Science, 312*, 1143-1144. Available: http://www.sciencemag.org/cgi/reprint/312/5777/1143.pdf [accessed October 2007].

Tran, L. (2009). *Children and adults' understanding of ocean and climate sciences.* Paper prepared for the Committee for the Review of the NOAA Education Program, National Research Council, Washington, DC. Available: http://www7.nationalacademies.org/bose/NOAA%20Teaching%20and%20Learning.pdf [accessed October 2009].

Tretter, T.R., Jones, M.G., Andre, T., Negishi, A., and Minogue, J. (2006). Conceptual boundaries and distances: Students' and experts' concepts of the scale of scientific phenomena. *Journal of Research in Science Teaching, 43*, 282-319.

Tytler, R. (2000). A comparison of year 1 and year 6 students' conceptions of evaporation and condensation: Dimensions of conceptual progression. *International Journal of Science Education, 22*(5), 447-467.

Tytler, R., and Peterson, S. (2000). Deconstructing learning in science: Young children's responses to a classroom sequence on evaporation. *Research in Science Education, 30*(4), 339-355.

U.S. Census Bureau. (2008). *An older and more diverse nation by midcentury.* Available: http://www.census.gov/Press-Release/www/releases/archives/population/012496.html [accessed July 2009].

U.S. Commission on Ocean Policy. (2004). *An ocean blueprint for the 21st century: Final report.* Washington, DC. Available: http://oceancommission.gov/documents/full_color_rpt/welcome.html [accessed February 2010].

U.S. Department of Commerce and U.S. Department of Education. (2008). *The shortage in the number of individuals with post-baccalaureate degrees in subjects related to fishery science.* Available: http://www.st.nmfs.noaa.gov/report_congress/ShortageOfDegrees.pdf [accessed October 2009].

U.S. Department of Education. (2004). *The Elementary and Secondary Education Act: The No Child Left Behind Act of 2001.* Available: http://www.ed.gov/policy/elsec/leg/esea02/index.html [accessed October 2009].

U.S. Department of Education. (2007). *Report of the Academic Competitiveness Council.* Washington, DC: Author. Available: http://www.ed.gov/about/inits/ed/competitiveness/acc-mathscience/report.pdf [accessed October 2009].

U.S. Department of Education. (2008). *Revenues for public elementary and secondary schools, by source of funds: Selected years, 1919-1920 through 2005-2006.* Available: http://nces.ed.gov/programs/digest/d08/tables/dt08_171.asp [accessed October 2009].

Uthe, R.E. (2000). Projecting the scientific method. *Science Teacher, 67*(9), 44-47.

Windschitl, M., Thomson, J., and Braaten, M. (2008). Beyond the scientific method: Model-based inquiry as a new paradigm of preference for school science investigations. *Science Education, 92*(5), 941-967.

Zacharia, Z.C., and Constantinou, C.P. (2008). Comparing the influence of physical and virtual manipulatives in the context of the Physics by Inquiry curriculum: The case of undergraduate students' conceptual understanding of heat and temperature. *American Journal of Physics, 76*(4), 425-430.

Appendix A

Committee Meeting Public Agendas

FIRST COMMITTEE MEETING

May 19-20, 2008

May 19, 2008

CLOSED SESSION
8:30-10:30 a.m.

OPEN SESSION

10:45 a.m.	**Public Welcome** *John Farrington,* Committee Chair, Scientist Emeritus, Woods Hole Oceanographic Institution *Michael Feder,* Study Director, Board on Science Education

NOAA Presentations and Discussions with the Committee

11:00 a.m.	**NOAA Office of Education Overview Structure and Role** **Description of the Major Education Programs** **Discussion of the Study Charge** *Louisa Koch,* Director, NOAA Office of Education

12:00 p.m. **Discussion of NOAA Presentation**
Discussion Moderator: *George Matsumoto,* Committee Member

12:30 p.m. **Continued Discussion** (lunch provided)

Panel Discussion: Ocean Policy Reports and Interagency Panels

1:30 p.m. **National Policy Reports and Interagency Panels**
James M. Coleman, Louisiana State University, U.S. Commission on Ocean Policy
Jeffery Reutter, Ohio State University, Ocean Research and Resources Advisory Panel
Lisa Rom, National Science Foundation; Cochair, NSTC Joint Subcommittee on Ocean Science and Technology
Sue Cook, Education Director, Consortium for Ocean Leadership (formerly CORE and JOI)

2:30 p.m. **Committee Discussion**
Discussion Moderator: *Frank Muller-Karger,* Committee Member, University of Massachusetts, Dartmouth

3:15 p.m. **Break**

CLOSED SESSION
3:30-8:00 p.m.

May 20, 2008

CLOSED SESSION
8:30-9:00 a.m.

OPEN SESSION
10:30 a.m. **Education Components of the National Sea Grant College Program**
Sharon Walker, Education Director, National Sea Grant College Program

11:15 a.m. **Committee Discussion**
Discussion Moderator: *Laura Murray,* Committee Member

11:45 a.m. **Continued Committee Discussion** (lunch provided)

12:45 p.m. **Evaluation of the Education Components of the National Sea Grant Program**
Leon Cammen, Director, National Sea Grant College Program

1:15 p.m. **Committee Discussion**
Discussion Moderator: *Frances Lawrenz,* Committee Member

Possible follow-up questions:
Have there been any attempts to coordinate what type of evaluation data is collected by the education projects?
Have the results of the evaluations led to changes in the education projects?
How much funding is set aside for the evaluation?

1:45 p.m. **Break**

CLOSED SESSION
2:00-4:30 p.m.

SECOND COMMITTEE MEETING

September 15-16, 2008

September 15, 2008

CLOSED SESSION
8:30-10:15 a.m.

OPEN SESSION

Climate and Atmosphere Panel

10:15 a.m.	**Presentations** *Roberta Johnson,* Director, University Corporation for Atmospheric Research *Jim Brey,* Director, Education Program, American Meteorological Society *Frank Niepold,* Cochair, U.S. Climate Change Science Program, Education Interagency Working Group; Climate Education Coordinator, NOAA Climate Program Office *Jill Karsten,* Member, U.S. Climate Change Science Program, Education Interagency Working Group; Program Director, NSF Directorate for Geosciences Education and Diversity Discussion Moderator: *Kevin Kloesel,* Committee Member
11:15 a.m.	**Committee Discussion with Panel**
12:00 p.m.	**Continued Small Group Discussions** (lunch provided)

Earth Systems and Environmental Science Education Panel

1:00 p.m.	**Presentations** *Jim Sanders,* President, National Association of Marine Laboratories *Gerry Lieberman,* President, State Environmental Education Roundtable *Daniel Barstow,* Director, TERC Center for Earth and Space Science Education

Discussion Moderator: *Clarice Yentsch*, Committee Member

1:45 p.m. **Committee Discussion with Panel**

CLOSED SESSION
2:30-8:00 p.m.

September 16, 2008

CLOSED SESSION
8:00-10:30 a.m.

OPEN SESSION

Local NOAA Education Project Staff Presentations

10:30 a.m. **Presentations (15 minutes each)**

Claire Fackler, National Education Liaison, National Marine Sanctuary Program
Seaberry Nachbar, Program Manager, B-WET California with *Jim Neiss-Cortez*, SF Rocks, San Francisco State University
Glen Alexander, Education Coordinator, Padilla Bay National Estuarine Research Reserve
Eric Boldt, Warning Coordination Meteorologist, National Weather Service

Discussion Moderator: *Craig Strang,* Committee Member

11:30 a.m. **Committee Discussion with Presenters**

12:00 p.m. **Continued Small Group Discussions** (lunch provided)

CLOSED SESSION
1:00-4:00 p.m.

THIRD COMMITTEE MEETING

December 11-12, 2008

December 11, 2008

CLOSED SESSION
8:15-10:15 a.m.

OPEN SESSION

10:15 a.m. **Panel Discussion of the Broadening Participation Commissioned Paper Outline**

Roger Levine, American Institutes for Research
Eric Riggs, Purdue University
Deidra Gibson, Hampton University
Ramon Lopez, University of Texas, Arlington

Discussion Moderator: *Rajul Pandya,* Committee Member

11:15 a.m. **Committee Discussion with Panel**
Discussion Moderator: Rajul Pandya, Committee Member

12:00 p.m. **Continued Small Group Discussions** (lunch provided)

1:00 p.m. **Panel Presentation of NOAA's Education Partnership Program**

Jacqueline J. Rousseau, NOAA Educational Partnership Program
Larry Robinson, Florida A&M University
Reza Khanbilvardi, City University of New York

1:20 p.m. **Committee Discussion with Panel**
Discussion Moderator: *Brett Moulding,* Committee Member

CLOSED SESSION
2:00-8:00 p.m.

December 12, 2008

CLOSED SESSION
8:30-10:30 a.m.

OPEN SESSION

	NOAA Education Program Presentations
10:30 a.m.	*Marci Wulff*, NOAA's Coral Reef Conservation Program *Ron Gird*, NOAA's National Weather Service *Atziri Ibanez* and *Laurie McGilvray*, National Estuarine Research Reserve System
11:30 a.m.	**Committee Discussion** Discussion Moderator: *Kim Kastens*, Committee Member
12:15 p.m.	**Continued Small Group Discussions** (lunch provided)

CLOSED SESSION
1:15-4:00 p.m.

FOURTH COMMITTEE MEETING

March 23-24, 2009

March 23, 2009

CLOSED SESSION
8:00-11:15 a.m.

OPEN SESSION

Evaluation and Implementation Presentations

11:15 a.m.	**Bennett TOPS Evaluation Model Implementation at NOAA** *Martin Storksdieck,* Institute Learning for Innovation
	Committee Discussion of TOPS Implementation Discussion Moderator: *Frances Lawrenz,* Committee Member
12:00 p.m.	**Continued Discussions of Evaluation** (lunch provided)
1:00 p.m.	**NOAA Education Implementation Plan** *Christos Michalopoulos,* NOAA
1:15 p.m.	**Committee Q&A on the Education Implementation Plan** *Laura Murray,* Committee Member
1:45 p.m.	**NOAA Education Projects Presentations**
	National Marine Sanctuaries Program (NMSP) *Michiko Martin,* National Education Coordinator NMSP
	Committee Discussion, *Craig Strang,* Committee Member
	National Marine Fisheries Services (NMFS) *Molly Harrison,* National Education Coordinator NMFS
	Committee Discussion, *Brett Moulding,* Committee Member

Ocean Explorer
Paula Keener-Chavis, Director, Education Programs, NOAA Ocean Exploration and Research Program

Committee Discussion, *James Coleman*, Committee Member

3:00 p.m. **Crosscutting Discussion of Programs**
James Coleman, Committee Member

CLOSED SESSION
3:30-8:30 p.m.

March 24, 2009

CLOSED SESSION
8:30 a.m.-4:15 p.m.

Appendix B

Site Visit Agendas

CHESAPEAKE BAY SMITHSONIAN ENVIRONMENTAL RESEARCH CENTER

June 8, 2009

NOAA Participants:	Doug Levin, Kevin Schabow, Shannon Sprague
Committee Members:	Gordon Kingsley, Laura Martin
Other Participants:	Tom Ackerman, Chesapeake Bay Foundation Jamie Baxter, Chesapeake Bay Trust Elena Takaki, Maryland Department of Natural Resources

9:00 a.m.	Introduction
9:30 a.m.	Office and Program Overview Shannon Sprague
9:45 a.m.	Emerging Scientist Program Kevin Schabow *Distance Learning*
10:15 a.m.	Observations in the K-12 Classroom Doug Levin *Build-A-Buoy* *BOBs (Basic Observation Buoys)*

	Chesapeake Bay Interpretive Buoy System Capabilities and Curriculum
11:30 a.m.	Exploring Our Oceans and Bays Doug Levin *Aquabots and AUVs*
12:00 p.m.	Lunch *Visit Bay Commitment and Gather BOB data*
1:00 p.m.	Travel to National Chesapeake Bay Office
1:15 p.m.	B-WET Shannon Sprague and partners

MARYLAND SEA GRANT AT SOUTH CARROLL HIGH SCHOOL

June 9, 2009 (morning)

NOAA Participant:	Adam Frederick
Committee Member:	Gordon Kingsley
Other Participants:	Marty, South Carroll High School student Judy Plaskowitz, South Carroll High School teacher Brad Yohe, Carroll County Public Schools
9:00 a.m.	South Carroll High School *Aquaculture Research Lab/Classroom*
11:00 a.m.	Center of Marine Biotechnology *Aquaculture Research Center*
12:20 p.m.	Lunch

CHESAPEAKE BAY NATIONAL ESTUARINE RESEARCH RESERVE SYSTEM AT OTTER POINT CREEK

June 9, 2009 (afternoon)

NOAA Participant:	Bart Merrick
Committee Member:	Gordon Kingsley

1:00 p.m.	Overview and Tour, Anita C. Leight Estuary Center
2:00 p.m.	Estuary Boat Tour and Discussion of Education Projects

MONTEREY BAY NATIONAL MARINE SANCTUARY

July 7, 2009

NOAA Participants:	Lisa Emanuelson, Dawn Hayes, Sacha Lozano, Paul Michel, Seaberry Nachbar
Committee Members:	George Matsumoto, Frances Lawrenz, Clarice Yentsch
8:30 a.m.	Introduction *Welcome by Superintendent Paul Michel*
9:00 a.m.	Overview of Day Introduction to Sanctuary Programs of the Region • School Programs • Teacher Programs and Curricula • Multicultural Programs (MERITO and Ocean for Life) • Public outreach on resource issues • Field Programming
10:30 a.m.	B-WET Field Experience Carmel Valley
12:00 p.m.	Lunch
12:30 p.m.	Team OCEAN Overview and Demonstration
2:30 p.m.	Wrap-Up

ELKHORN SLOUGH NATIONAL ESTUARINE RESEARCH RESERVE

July 8, 2009

9:00 a.m.	Overview of Programs and Plans for the Day
9:30 a.m.	Tour and Discussion of Visitor Center Exhibits
10:00 a.m.	Estuary Hiking Tour/Field Experience
12:30 p.m.	Lunch
1:00 p.m.	Observation of the Teaching Lab
2:00 p.m.	Discuss Programs and Plans for the Future
3:00 p.m.	Departure

Appendix C

NOAA Education Program Websites

Education Program	Website
Bay-Watershed Education and Training	http://www.oesd.noaa.gov/BWET/
Climate Program	http://www.climate.noaa.gov/education/
Cooperative institutes	http://www.nrc.noaa.gov/ci/
Coral Reef Conservation Program	http://coralreef.noaa.gov/education/
Educational Partnership Program	http://www.epp.noaa.gov/
Environmental Literacy Grants	http://www.oesd.noaa.gov/elg_faqs.html
Ernest F. Hollings Scholarship Program	http://www.oesd.noaa.gov/Hollings_info.html
Dr. Nancy Foster Scholarship Program	http://fosterscholars.noaa.gov/
JASON	http://www.jason.org
National Estuarine Research Reserve System	http://nerrs.noaa.gov/Education/.aspx
National Ocean Service	http://oceanservice.NOAA.gov/education/
National Marine Sanctuaries	http://sanctuaries.noaa.gov/education/welcome.html
Ocean Explorer	http://oceanexplorer.noaa.gov/edu/welcome.html
Ocean Hall	http://www.noaanews.noaa.gov/stories2008/20080925_oceanhall.html
Sea Grant	http://www.seagrant.noaa.gov/roe/education.html
StormReady/TsunamiReady	http://www.stormready.noaa.gov/ http://www.tsunamiready.noaa.gov/
Teacher at Sea	http://teacheratsea.noaa.gov/

Appendix D

Biographical Sketches of Committee Members and Staff

John W. Farrington (*Chair*) is interim dean and professor in the School of Marine Science and Technology at the University of Massachusetts, Dartmouth, and scientist emeritus at the Woods Hole Oceanographic Institution (WHOI) with expertise in marine chemistry and geochemistry. He joined WHOI in 1971 as a postdoctoral investigator. He held successive positions in the chemistry department for 17 years and simultaneously served for six years as director of the WHOI Coastal Research Center. In 1988 he was appointed Michael P. Walsh professor and director of the Environmental Sciences Program at the University of Massachusetts, Boston. In 1990, he returned to WHOI to become associate director for education and dean of graduate studies. In 2002, Farrington was named vice president for academic programs and dean at WHOI. His research interests include marine organic geochemistry, the biogeochemistry of organic chemicals of environmental concern, ocean science education, and the interaction between science and policy. He has served on committees and panels for international, national, and local organizations, including the UNESCO-Intergovernmental Oceanographic Commission, the National Academy of Sciences, the National Science Foundation, the Office of Naval Research, the Commonwealth of Massachusetts, and the Lloyd Center for Environmental Studies. At the National Research Council, he has participated in seven consensus studies, chairing three of them, and has been a member of the Environmental Studies Board, the Board on Environmental Studies and Toxicology, and the Marine Board. He has B.S. and M.S. degrees in chemistry and a Ph.D. in oceanography from the University of Rhode Island.

James M. Coleman is Boyd professor at the Coastal Studies Institute of Louisiana State University and Agricultural and Mechanical College. His research interests include coastal and marine processes and coastal management. The training of scientists and engineers to compete in a technological, global environment is central to his ongoing areas of research. At the National Research Council, he chaired the Marine Board and served as a member of the Ocean Studies Board. He is a member of the National Academy of Engineering and the Russian Academy of Natural Sciences. He has received many awards in his nearly 40-year scientific career, including the Kapitsa Medal of Honor for his contributions to the field of petroleum sciences. He has B.S., M.S., and Ph.D. degrees in geology from Louisiana State University and Agricultural and Mechanical College.

Michael A. Feder (*Senior Program Officer*) is staff member with the Board on Science Education. He is the director the Review of NOAA Education Programs and the Roundtable on Climate Change Education, and is working on study to develop a Conceptual Framework for New Science Education Standards and the review of Discipline Based Education Research. Previously, he staffed the National Research Council study of Learning Science in Informal Environments, the study of K-12 Engineering Education, and the review of NASA's pre-college education programs. Prior to joining the National Research Council he conducted evaluations of and provided technical assistance for national, state, and local education efforts. He has an M.A. and a Ph.D. in Applied Developmental Psychology from George Mason University.

Janet Hustler is coprincipal investigator and former director of the Partnership for Student Success in Science (PS3), a Math-Science Partnership project funded by the National Science Foundation. Located at San Jose State University, PS3 features science professional development for middle school teachers. The project provides teachers with professional development to enrich their content background and pedagogical skills, along with critical friends study groups to help teachers infuse what they learn into their practice. It also offers teacher leadership training plus coaching and summer institutes. Prior to her work at PS3, Hustler was the principal investigator and director of a similar six-year project focusing on elementary science teaching and learning. Her background includes more than 20 years of classroom teaching experience and out-of-the-classroom roles, such as science coordinator for the Palo Alto Unified School District. She has served on several national boards, including the Leadership and Assistance in Science Education Reform (LASER) and the Association of Science Materials Centers board, and has been a faculty member of numerous LASER institutes. At the National Research Council, she is a member of the Board on Science Education. Hustler has an M.A. in social sciences from San Jose State University, an M.A. in educational administration from Santa Clara University, and an M.S. in marine science education from Oregon State University.

Kim A. Kastens is a Doherty senior research scientist at the Lamont-Doherty Earth Observatory of Columbia University and an adjunct professor of earth and environmental sciences. Her training and early career were in marine geology, focusing on the geological evolution of the Mediterranean region and the structure and tectonics of transform faults. Over the past 15 years, her focus has shifted toward improving public understanding of the Earth and the environment, through training of environmental journalists, professional development of teachers, innovative use of information technology, and research on the science of learning. Her educational efforts have included founding Columbia's dual master's degree program, Earth and Environmental Science Journalism, developing the *Where Are We?* software to help children learn to read maps, and developing Data Puzzles to foster use of authentic geoscience data in high schools. Her research on learning projects investigate how children use maps while navigating, how climate forecast maps and bathymetric maps are understood by their target audiences, and how people visualize a three-dimensional geological structure from the limited information available from outcrops. She is currently co-leading a multidisciplinary effort to create a Synthesis of Research on Thinking and Learning in the Geosciences. Kastens has a B.S. in geology and geophysics from Yale University and a Ph.D. in oceanography from the Scripps Institution of Oceanography, University of California, San Diego.

Gordon Kingsley is associate professor in the School of Public Policy at the Georgia Institute of Technology, where his teaching and research focus on science and technology policy and public management. His research examines the development and implementation of effective partnerships across the public, private, and nonprofit sectors. His current research projects explore the impacts of partnerships on the development and allocation of scientific and technical human capital. This work is being conducted in three policy domains, examining the impact of educational partnerships between universities and K-12 schools on the development of math and science instructors and instruction; strategies used by state transportation agencies for effectively managing large numbers of engineering consultants and contractors; and the development of hybrid organizations and network organizations designed to stimulate technology-led economic development. He has served as a consultant or researcher for the National Science Foundation, the Manufacturing Extension Partnership of the National Institute of Standards and Technology, the Office of Technology Policy in the U.S.

Department of Commerce, the Office of Science in the U.S. Department of Energy, a science advisory board of the U.S. Environmental Protection Agency, and state-level agencies. He has a B.A. in international affairs and economics from American University, an M.S. in international affairs from Columbia University, and a Ph.D. in public administration from Syracuse University.

Kevin Kloesel is associate dean for public service and outreach in the College of Atmospheric and Geographic Sciences at the University of Oklahoma. He is directly responsible for outreach programs and tours for the over 30,000 people who visit the National Weather Center facility in Norman annually. In addition, he is an associate professor in the Oklahoma University School of Meteorology, with teaching and research interests ranging from synoptic meteorology to societal impacts and decision making in weather-impacted situations. He led the team that won the Innovations in American Government Award from Harvard University and the Ford Foundation for their work with the emergency management community in Oklahoma. Currently, he works directly with thousands of K-12 students and teachers, as well as hundreds of emergency management agencies, in finding appropriate applications for weather data in local education and decision making. He was also a content designer for Scholastic Magazine's The Magic School Bus Kicks Up a Storm children's museum exhibit. Previously, he was director of outreach for the Oklahoma Climatological Survey and served as director of the Florida Climate Center in Tallahassee. While a tenured faculty member at Florida State University, he served as a research fellow with the Cooperative Institute for Tropical Meteorology and co-directed an outreach project, EXPLORES!, which provided NOAA satellite data to over 200 schools. He has a B.S. in engineering science from the University of Texas at Austin and M.S. and Ph.D. degrees in meteorology from the Pennsylvania State University.

Frances Lawrenz is associate vice president for research and professor of psychological foundations and quantitative methods in education at the University of Minnesota. She conducts research in science and mathematics program evaluation. Her evaluations use a variety of techniques and usually involve mixing quantitative and qualitative methods. She is currently involved in the evaluation of several national science and mathematics programs, including the Collaborative Evaluation Communities in Urban Schools, the Active Physics Curricular Development, and the Impact and Effectiveness of the Collaboratives for Excellence in Teacher Preparation Program. She has also completed a number of major evaluations, including the Systemic Initiative Evaluation, the Long Term Effects of Teacher Enhancement Evaluation, and the Authentic Assessment Systems for Constructivist Based Elementary Science Programs Evaluation. She is interested in instrument development and in distinguishing, among various types of assessments, those that are most appropriate for a given situation. She won the Graduate-Professional Teaching Award for her contributions to graduate and professional education in 2002. She applies evaluation methods to courses, programs, and advising to improve them. She has a B.S. in chemistry with a minor in mathematics, an M.A. in education, and a Ph.D. in education with related fields of study in chemistry and mathematics from the University of Minnesota, Minneapolis.

George I. Matsumoto is senior education and research specialist at the Monterey Bay Aquarium Research Institute. His role involves a variety of activities: seminar coordinator, summer internship coordinator, Livelink mentor, distance education, and links between the research institute and other partners. His research interests focus on deep sea communities, particularly invertebrates in the open ocean. Specific areas of interest include ecology and biogeography of open ocean and deep sea organisms; functional morphology, natural history, and behavior of pelagic and benthic organisms; and systematic and evolution of ctenophores and cnidarians (molecular phylogeny). Matsumoto is active in several public service efforts, including as a volunteer scientist for Bay Area Schools for Excellence in Education, education session chair for the U.S. Ocean Research Priorities Plan, and board member of the Friends of the Monterey Academy of Oceanographic Sciences and for Camp SEALab. At the National Research Council, he was a member of the committee on the evaluation of the Sea Grant program review process and is a member of the Ocean Studies Board. He has an A.B. in marine botany from the University of California, Berkeley, and a Ph.D. in marine biology from the University of California, Los Angeles.

Brett D. Moulding is the director of the Utah Partnership for Effective Science Teaching and Learning, a four district professional development collaborative. He was the director of curriculum and instruction at the Utah State Office of Education before retiring in January 2008. He was the state science education specialist and coordinator of curriculum from 1993 to 2004. At the National Research Council, Moulding is a member of the Board on Science Education. He taught chemistry for 20 years at Roy High School in the Weber District Science and served as the district teacher leader for 8 years. Moulding received the Governor's Teacher Recognition Award, the Presidential Award for Excellence in Mathematics and Science Teaching, and the Award of Excellence from the Governor's Science and Technology Commission. He served on the Triangle Coalitional Board and the NAEP 2009 Framework Planning Committee and was president of the Council of State Science Supervisors from 2003 to 2006. He has a B.S. in chemistry from the University of Utah and an M.Ed. from Weber State University. He has an administrative supervisory certificate from Utah State University.

Frank E. Muller-Karger is a professor of biological oceanography at the University of South Florida. He is a biological oceanographer who conducts research on marine primary production using satellite remote sensing, large data sets, networking, and high-speed computing. His present work focuses on assessing the importance of continental margins, including areas of upwelling, river discharge, and coral reefs in the global carbon budget, using satellites that measure ocean color and sea surface temperature. Muller-Karger has worked to educate K-12 teachers about the use of new technologies in oceanography through workshops sponsored by the National Aeronautical and Space Administration (NASA). He is interested in addressing the problem of underserved and underrepresented groups in academic science programs and has advocated for minorities, educators, and science education as a member of the U.S. Commission on Ocean Policy. He served as director of the Institute for Marine Remote Sensing at the University of South Florida and as the science adviser for the Florida Center for Ocean Science Education Excellence. At the National Research Council, he was a member of the Ocean Studies Board and has served on the committees for Extending Observations and Research Results to Practical Applications: A Review of NASA's Approach and An Assessment of Balance in NASA's Science Programs. He received the NASA Jet Propulsion Laboratory Award for Outstanding Contributions and the NASA Administrator Award for Exceptional Contribution and Service for supporting development of satellite technologies for ocean observation. He also received the Julius A. Stratton Award for Leadership. He has a B.S. in marine science from the Florida Institute of Technology, an M.S. in marine science from the University of Alaska, Fairbanks, and a Ph.D. in marine science from the University of Maryland.

Laura Murray is a research professor at the University of Maryland's Center for Environmental Science Horn Point Laboratory. Her expertise and research interests include seagrass and wetlands ecology, with a focus on the response of submerged aquatic vegetation to nutrient enrichment. As an educator, Murray's primary goal has been to link the world-class research with science education. Her involvement in environmental science education includes conducting professional development workshops for teachers and informal educators, providing research experiences for teachers, establishing research-based programs for K-12 students, and administrating programs that partner scientists, educators, and students. She is currently the director of the Center for Ocean Science Education Excellence Coastal Trends and the director of the Environmental Science Education Center at the Horn

Point Laboratory. She has published in both the scientific research and in the science education fields. Murray has a B.S. in marine science and an M.S.T. in biology/education from the University of West Florida and a Ph.D. in wetlands ecology from the College of William and Mary.

Rajul Pandya is the director of the Community Building Program at the University Corporation for Atmospheric Research (UCAR). Its mission is to build and support institutional relationships that will increase the diversity and societal relevance of the atmospheric and related sciences. Pandya is also director of SOARS, an internship program to broaden participation in the atmospheric and related sciences through research experience, mentoring, and a strong learning community. He also serves as coordinator of UCAR's Africa Initiative, which seeks to support atmospheric research and applications in West Africa through capacity building and collaborative research. Pandya's past scientific work has involved analytical and numerical modeling of convection and other atmospheric phenomena, and his teaching has focused on enabling students to learn by working directly with visualizations and data in a variety of settings. Pandya has a B.S. in physics from the University of Illinois, Urbana-Champaign, and a Ph.D. in atmospheric sciences from the University of Washington, Seattle.

Craig Strang is associate director of the Lawrence Hall of Science (LHS) at the University of California, Berkeley, where he leads the Center for Leadership in Science Teaching and the Center for Ocean Sciences Education Excellence-California (COSEE-CA). He was the first chair of the National COSEE Council. In addition, he is founding director of Marine Activities Resources and Education. Strang is the author of three multivolume sets of science and environmental education curriculum materials for grades K-8 and has developed professional development networks to support the implementation of each of these programs. Also, he authored three teacher guides published by the LHS Great Explorations in Math and Science program: *On Sandy Shores, Ocean Currents,* and *Only One Ocean.* He was the principal project consultant responsible for the creation and funding of a high school environmental justice internship program, XCEL: Cross-Cultural Environmental Leadership for Audubon Canyon Ranch. He is interested in the use of inquiry-based science instruction to promote language acquisition among English language learners. Strang has conducted field research on elephant seals and humpback whales and occasionally leads natural history ecotours to Baja California and Galapagos. He is past president of the Southwest Marine Educators Association and is a member of the executive committee of the board of directors of the National Marine Educators Association. He has a B.A. in environmental studies from the University of California, Santa Cruz.

Clarice Yentsch is adjunct research scientist at the Nova Southeastern University Oceanographic Center in Dania Beach, Florida. She is also a consultant for the National Museum of Natural History at the Smithsonian Institution in Washington, DC. Previously, she was a research scientist and educator at the Mel Fisher Maritime Museum. Yentsch is the co-founder of Bigelow Laboratory for Ocean Sciences in West Boothbay Harbor, Maine, and is founder of the J.J. MacIsaac Flow Cytometery/Cell Sorting Facility there. She studies dinoflagellates, which cause toxic red tides and are symbiotic in reef-building corals, flow cytometry, and cell sorting. She is responsible for system reforms and curriculum development at the Education Development Center in Newton, Massachusetts. She has served on professional advisory boards and board of trustees for profit and not-for-profit science organizations. From 1998 to 2002, Yentsch served as an independent consultant with the American Museum of Natural History in New York. She has a B.S. in natural science and an M.A.T. in education from the University of Wisconsin, Madison, and a Ph.D. in oceanography from Nova Southeastern University.